Strength of Materials

K. Bhaskar · T. K. Varadan

Strength of Materials

A Concise Textbook

K. Bhaskar
Department of Aerospace Engineering
Indian Institute of Technology Madras
Chennai, India

T. K. Varadan
(Emeritus) Department of Aerospace Engineering
Indian Institute of Technology Madras
Chennai, India

ISBN 978-3-031-06376-3 ISBN 978-3-031-06377-0 (eBook)
https://doi.org/10.1007/978-3-031-06377-0

Jointly published with Ane Books Pvt. Ltd.
In addition to this printed edition, there is a local printed edition of this work available via Ane Books in South Asia (India, Pakistan, Sri Lanka, Bangladesh, Nepal and Bhutan) and Africa (all countries in the African subcontinent).
ISBN of the Co-Publisher's edition: 978-93-90658-43-5

This Springer imprint is published by the registered company Springer Nature Switzerland AG
The registered company address is: Gewerbestrasse 11, 6330 Cham, Switzerland

Preface

(Addressed to students)

As teachers who have taught this subject for decades, we know that many of you consider it a little tricky if not altogether difficult. The aim of this book, intentionally kept slim, is to allay such fears and to show that a quick grasp of the essentials and a systematic approach to problem-solving would help you enjoy studying this content. All the topics normally covered in a corresponding first-level course are dealt with here in sufficient detail.

The book is expected to be useful as a gentle introduction to this important subject and also as a quick refresher in your career later.

Chennai, India

K. Bhaskar
T. K. Varadan

Contents

About the Authors

Prof. K. Bhaskar has been with the Department of Aerospace Engineering, Indian Institute of Technology, Madras, since 1992. As part of the Structures Group, he teaches courses related to Solid Mechanics and Elasticity. His research contributions are primarily related to theoretical modeling of thick laminated structures, with around fifty publications in refereed journals.

Prof. T. K. Varadan was with the Department of Aerospace Engineering, Indian Institute of Technology, Madras, for about thirty-five years before retiring in 2001. Besides teaching a wide variety of courses related to Structural Mechanics and Aircraft Design, he has made significant research contributions in the areas of Nonlinear Vibrations and Composite Structures, with more than one hundred refereed publications.

1 Preliminaries

Discussion in this book is confined to structures

- that are isotropic (i.e. material properties at any point same in all directions) and homogeneous (i.e. material properties same at all points of the structure);
- that are statically loaded such that displacements and rotations are small (see Sect. 1.3);
- that exhibit linear elastic behaviour (i.e. the loading is such that the internal stresses are within the elastic limit and the structure regains its original shape and size after unloading; further, a linear constitutive law is applicable—see Sect. 1.6).

1.1 State of Stress at a Point

For a three-dimensional body in equilibrium under the action of applied loads as well as support reactions, the state of stress at any interior point P may be specified in terms of nine components—three normal stresses (σ_{xx}, σ_{yy}, σ_{zz}) and three pairs of shear stresses (τ_{yz} & τ_{zy}, τ_{zx} & τ_{xz}, τ_{xy} & τ_{yx}) defined as follows with reference to any convenient Cartesian x, y, z coordinates. Assume a cutting plane perpendicular to x-axis and passing through P. Separating the two free bodies (Fig. 1.1) and knowing that a distribution of internal forces should be transmitted through the interface for equilibrium, consider an infinitesimal area ΔA around P; let the force transmitted through this area be ΔF, which, in general, will be in some direction inclined to x, y as well as z. Resolving ΔF along the axes as ΔF_x, ΔF_y, ΔF_z, three stress components may be defined at P as

$$\sigma_{xx} = \lim_{\Delta A \to 0} \frac{\Delta F_x}{\Delta A}, \quad \tau_{xy} = \lim_{\Delta A \to 0} \frac{\Delta F_y}{\Delta A}, \quad \tau_{xz} = \lim_{\Delta A \to 0} \frac{\Delta F_z}{\Delta A}$$

K. Bhaskar and T. K. Varadan, *Strength of Materials*,
https://doi.org/10.1007/978-3-031-06377-0_1

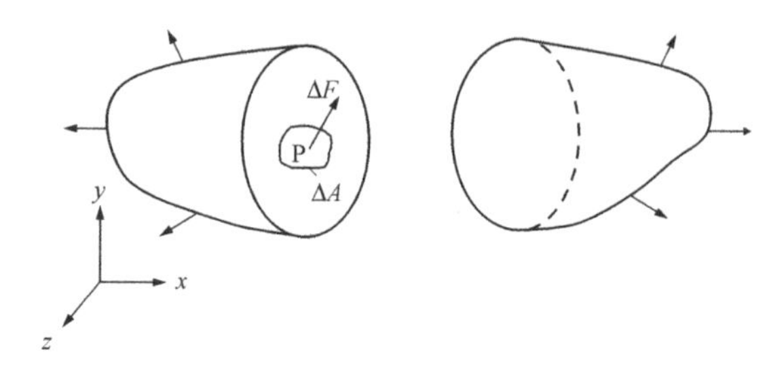

Fig. 1.1 Definition of σ_{xx}, τ_{xy}, τ_{xz}

where the first subscript indicates the normal to the cutting plane and the second the direction in which the stress acts; thus σ_{xx} is the normal stress and τ_{xy}, τ_{xz} are the tangential or shear stresses with reference to this cutting plane. Similarly, the remaining stress components are defined by considering cutting planes perpendicular to y and z and passing through P.

All these components may be conveniently shown on an infinitesimal cubical element at P as in Fig. 1.2 where the left face of the element represents the area around P on the left face of the right free body of Fig. 1.1, and the right face that on the right face of the left free body; the other faces of the element may similarly be associated with the free bodies generated by cutting planes perpendicular to y and z.

The positive sign convention for the stress components is also shown in Fig. 1.2; on faces with outward normal along positive x or y or z, the stresses are positive when acting along the positive coordinate directions, and vice versa. This implies that normal stresses are positive when tensile, and negative when compressive.

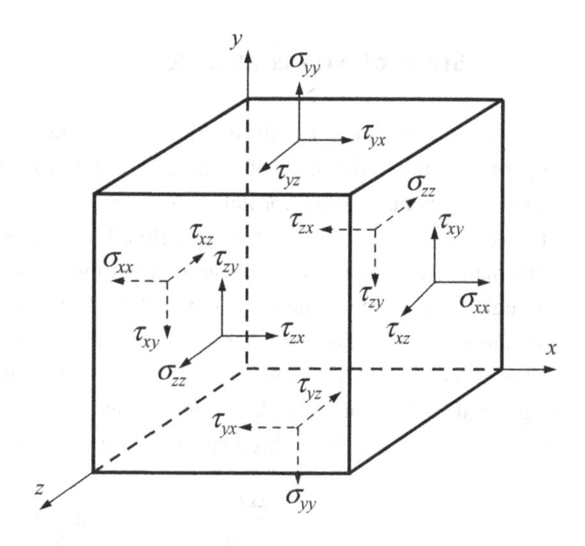

Fig. 1.2 Stress components

1.2 Principle of Complementary Shear

Considering the equilibrium of the element of Fig. 1.2, it is clear that force equilibrium in any direction is automatically satisfied while moment equilibrium will be satisfied only if $\tau_{ij} = \tau_{ji}$ for $i, j = x, y, z$. Hence, τ_{xy}, and τ_{yx} of the same algebraic sign and magnitude are complementary to each other at any point, and so on.

Thus, at any point, there are only six independent stress components—σ_{xx}, σ_{yy}, σ_{zz} (or σ_x, σ_y, σ_z in shortened form), and τ_{yz}, τ_{zx}, τ_{xy}, with the units of N/m^2 (also called pascals—Pa).

1.3 The Assumption of Small Displacements and Rotations

This assumption is required so as to avoid inconvenient nonlinear equations in the final theories to be developed and to enable equilibrium equations to be derived with reference to the undeformed geometry itself.

The assumption may be stated thus: "the displacement of any point of the structure due to applied loading is small with respect to the smallest linear dimension of the structure, and the corresponding angle of rotation (in radians) of any linear fiber of the structure is small compared to unity." It can also be stated more formally as: "the strain components are all small compared to unity, and displacement derivatives are small such that their squares and products may be neglected in the strain–displacement relations."

The above assumption is in general valid for structures undergoing elastic deformation; however, in some problems involving flexible structures such as long, slender beams or very thin plates, some non-linear terms may have to be retained in the strain–displacement relations even when the structure is loaded within the elastic limit. Even in such cases, a solution of the linear problem is often required as a first step.

An important advantageous feature of linear analysis is the validity of the superposition principle; this enables the consideration of just one applied load at a time, and thus decomposition of a problem of combined loading into simpler problems which may be analysed individually as that of a bar or shaft, etc., and the direct use of tabulated results from handbooks for elementary load cases.

At this juncture, it should be noted that all the displacements as shown in various figures of this book are highly exaggerated for the sake of clarity and are actually quite small.

1.4 State of Strain at a Point

At any point, the deformation of an infinitesimal cubical element (Fig. 1.3a) due to normal stresses alone is such that the element becomes a cuboid (Fig.1.3b) with all sides changed

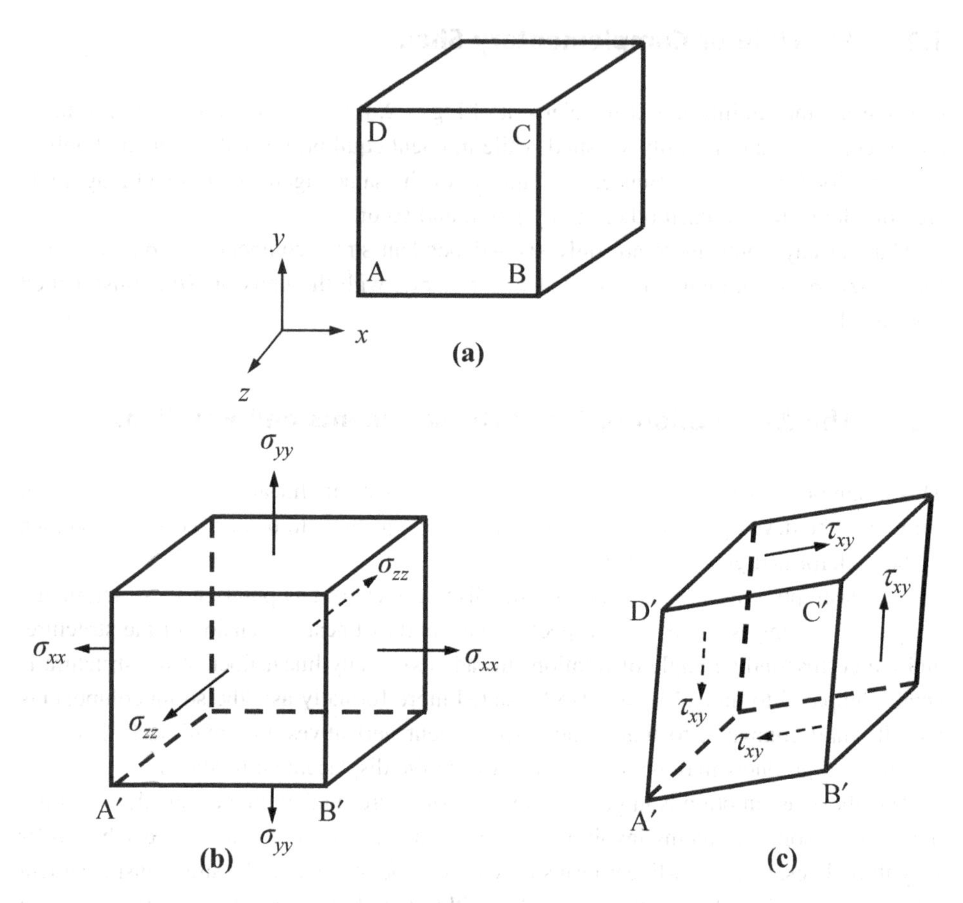

Fig. 1.3 (**a**) Undeformed geometry; (**b**) deformation due to normal stresses; (**c**) deformation due to τ_{xy}

in length. The normal strains correspond to such fractional changes in length and are defined as

$$\varepsilon_{xx} = \lim_{AB \to 0} \frac{A'B' - AB}{AB}$$

and similarly, ε_{yy} and ε_{zz}.

Due to any one shear stress component, say τ_{xy} alone, the cubical element undergoes distortion wherein the front and rear faces take the shape of a rhombus with lengths of the sides unchanged while the other faces simply rotate without a change of shape or size (Fig. 1.3c). The departure from the right angle between lines originally along x and y directions is defined as engineering shear strain γ_{xy}

$$\text{i.e.}\quad \gamma_{xy} = \frac{\pi}{2} - \angle B'A'D' = \angle A'D'C' - \frac{\pi}{2}\ \text{(in radians)}$$

Similarly, γ_{yz} and γ_{zx} may be defined as the distortions of the right (or left) and top (or bottom) faces occurring due to τ_{yz} and τ_{zx}, respectively. Thus, we have six strain components—ε_{xx}, ε_{yy}, ε_{zz} (or ε_x, ε_y, ε_z in shortened form), and γ_{yz}, γ_{zx}, γ_{xy}—all of them being dimensionless.

The sign convention for strains is that extensional normal strains are positive while positive shear strains correspond to distortions caused by the corresponding positive shear stresses; thus, for example, γ_{xy} as shown in Fig. 1.3c is positive.

1.5 Strain–Displacement Relations

If u, v, w are the x, y, z displacement components, respectively, then

$$\varepsilon_x = \frac{\partial u}{\partial x},\quad \varepsilon_y = \frac{\partial v}{\partial y},\quad \varepsilon_z = \frac{\partial w}{\partial z}$$

$$\gamma_{yz} = \frac{\partial v}{\partial z} + \frac{\partial w}{\partial y},\quad \gamma_{xz} = \frac{\partial u}{\partial z} + \frac{\partial w}{\partial x},\quad \gamma_{xy} = \frac{\partial u}{\partial y} + \frac{\partial v}{\partial x}$$

1.6 Linear Constitutive Law

Considering the deformation due to uniaxial stress, say σ_x alone, we have, by simple Hooke's law, a direct strain given by $\varepsilon_x = (\sigma_x/E)$ where E is the Young's modulus. Due to Poisson effect, lateral strains are also caused by σ_x as given by

$$\varepsilon_y = \varepsilon_z = -\mu\left(\frac{\sigma_x}{E}\right)$$

where μ is the Poisson's ratio.

If σ_y and σ_z are also present, they also cause direct and lateral strains similar to the above, so that the net effect of all the three normal stresses can be written as

$$\varepsilon_x = \frac{\sigma_x - \mu(\sigma_y + \sigma_z)}{E},\quad \varepsilon_y = \frac{\sigma_y - \mu(\sigma_x + \sigma_z)}{E},\quad \varepsilon_z = \frac{\sigma_z - \mu(\sigma_x + \sigma_y)}{E}$$

The presence of shear stresses does not affect the above normal strains.

As noted earlier, any shear stress component causes the corresponding shear strain alone; they are related to each other by the shear modulus or modulus of rigidity G, as given by

$$\gamma_{ij} = \frac{\tau_{ij}}{G} \text{ for } i \neq j, \quad i, j = x, y, z$$

The above shear strains are not affected by the presence of any of the normal stresses. Thus, the above two sets of equations give the net strains due to a general state of stress.

The equations for normal strains may be inverted and expressed conveniently as

$$\sigma_i = 2G\varepsilon_i + \lambda e \quad \text{for } i = x, y, z$$

wherein e is the volumetric strain or fractional change in volume ($\Delta V/V$) given by $(\varepsilon_x + \varepsilon_y + \varepsilon_z)$, and G and λ are called Lamé constants; they are related to E and μ as

$$G = \frac{E}{2(1+\mu)}; \quad \lambda = \frac{\mu E}{(1+\mu)(1-2\mu)}$$

An additional elastic constant called bulk modulus K is defined as the ratio of the hydrostatic pressure p (i.e. $\sigma_x = \sigma_y = \sigma_z = -p$) to the resulting fractional decrease in volume of a small element; this is related to E and μ as

$$K = \frac{E}{3(1-2\mu)}$$

Note that only two of the above constants (E, μ, G, λ, K) are independent.

Since E, G, and K are always positive, it follows that μ should be within the interval $(-1, ½)$; in practice, negative values are rare. For most metals, $\mu \approx 0.3$.

1.7 Plane Stress and Plane Strain

These are special two-dimensional cases (say, in x–y plane) wherein the stresses σ_x, σ_y, τ_{xy} and the corresponding strains do not vary in the transverse z-direction.

For plane stress, $\sigma_z = \tau_{xz} = \tau_{yz} = 0$, and this is taken to be applicable when the transverse dimension is small, and the thin, flat structure is subjected to in-plane loads alone; a simple example is a thin rotating disk. For this case, the in-plane stresses are related to the in-plane strains as

$$\begin{Bmatrix} \sigma_x \\ \sigma_y \end{Bmatrix} = \frac{E}{(1-\mu^2)} \begin{bmatrix} 1 & \mu \\ \mu & 1 \end{bmatrix} \begin{Bmatrix} \varepsilon_x \\ \varepsilon_y \end{Bmatrix}; \quad \tau_{xy} = G\gamma_{xy}$$

For plane strain, $\varepsilon_z = \gamma_{xz} = \gamma_{yz} = 0$, and this is applicable when the transverse dimension is large, and the long structure is prismatic, restrained at the ends, and subjected to loads that do not vary along the length; a simple example is a long pipe fixed at the ends and subjected to internal fluid pressure. For this case, the stress–strain law is given by

$$\sigma_i = 2G\varepsilon_i + \lambda(\varepsilon_x + \varepsilon_y) \text{ for } i = x, y; \quad \tau_{xy} = G\gamma_{xy}$$

Note that $\varepsilon_z = -\frac{\mu(\sigma_x + \sigma_y)}{E} \neq 0$ for plane stress,

and $\sigma_z = \mu(\sigma_x + \sigma_y) \neq 0$ for plane strain.

1.8 Two-Dimensional Stress Transformation

For plane stress or plane strain, the transformation rules yielding stress components at a point with respect to rotated x'–y' axes in terms of those with respect to x–y axes (Fig. 1.4) can be concisely expressed as

$$\sigma_{ij} = \sum_k \sum_l a_{ik} a_{jl} \sigma_{kl}, \quad i, j = x', y', \text{ and } k, l = x, y$$

where a_{rs} denotes the direction cosine of r with respect to s, and $\sigma_{rs} = \tau_{rs}$ when $r \neq s$.

For example,

$$\begin{aligned} \tau_{x'y'} &= a_{x'x} a_{y'x} \sigma_{xx} + a_{x'x} a_{y'y} \tau_{xy} + a_{x'y} a_{y'x} \tau_{yx} + a_{x'y} a_{y'y} \sigma_{yy} \\ &= -(\sigma_x - \sigma_y) \sin\theta \cos\theta + \tau_{xy}(\cos^2\theta - \sin^2\theta) \end{aligned}$$

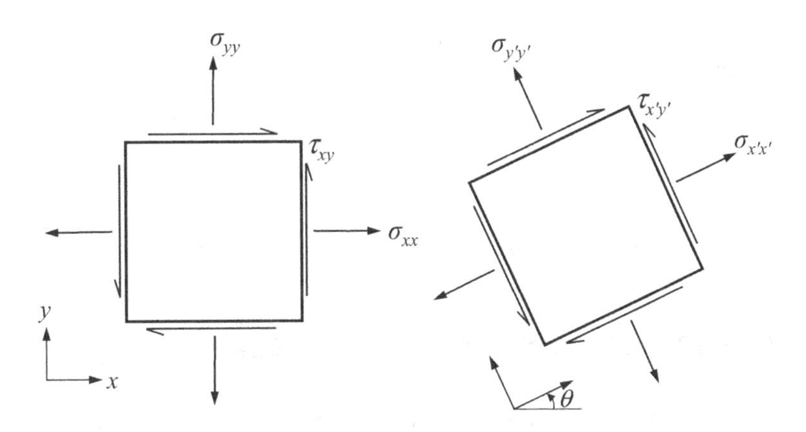

Fig. 1.4 Stress transformation

1.9 Principal Planes and Principal Stresses

Using the above 2-D stress-transformation rules, it can be shown that the normal stress $\sigma_{x'x'}$ becomes a maximum or minimum on planes corresponding to

$$\tan 2\theta = \frac{2\tau_{xy}}{\left(\sigma_x - \sigma_y\right)}$$

with one value of θ in the interval $(-\pi/4, \pi/4)$ and the other obtained by adding $\pi/2$. These orthogonal planes are referred to as principal planes; the corresponding extreme values of the normal stress are the principal stresses and are given by

$$\sigma_{P_1}, \sigma_{P_2} = \frac{\sigma_x + \sigma_y}{2} \pm \sqrt{\left(\frac{\sigma_x - \sigma_y}{2}\right)^2 + \tau_{xy}^2}$$

The shear stress turns out to be zero on these principal planes.

It should be added that the x–y plane itself is a shear-free, principal plane for plane stress or plane strain problems considered here, and the corresponding normal stress σ_z, having a value of 0 or $\mu(\sigma_x + \sigma_y)$, respectively, is referred to as the third principal stress denoted by σ_{P_3}. (For the more general three-dimensional state of stress, the three principal stresses and their orientations have to be found out using three-dimensional transformation rules.)

1.10 Maximum In-Plane Shear Stress

For plane stress or plane strain, the shear stress $\tau_{x'y'}$ becomes a maximum on planes at 45° to the principal planes and is given by

$$\left|\tau_{x'y'\max}\right| = \left|\frac{\sigma_{P_1} - \sigma_{P_2}}{2}\right| = \left|\sqrt{\left(\frac{\sigma_x - \sigma_y}{2}\right)^2 + \tau_{xy}^2}\right|$$

On these planes, there is a nonzero normal stress given by $(\sigma_x + \sigma_y)/2$.

1.11 Strain Transformation and Principal Strains

The equations governing strain transformation and principal strain analysis are identical to those for stresses with normal stresses replaced by the corresponding normal strains and shear stresses replaced by their "tensorial shear strain" counterparts ε_{ij} defined as

$$\varepsilon_{ij} = \frac{\gamma_{ij}}{2}$$

Note that planes of principal stresses and those of principal strains are coincident.

1.12 Mohr's Circle

This is a convenient semi-graphical aid for transformation of stresses or strains; it is explained with reference to illustrative examples in Sect. 8.2.2.

1.13 Notable Points About 2-D Transformation

- $\sigma_{x'} + \sigma_{y'} = \sigma_x + \sigma_y = \sigma_{P_1} + \sigma_{P_2}$, and similarly for strains.
- If $\sigma_x = -\sigma_y$ along with $\tau_{xy} = 0$ (i.e. with x, y axes being the principal axes), then $\left|\tau_{x'y'\max}\right| = |\sigma_x|$ with no normal stress on these maximum shear planes; if $\tau_{xy} \neq 0$, the principal stresses will be related to each other as $\sigma_{P_1} = -\sigma_{P_2}$, and $\left|\tau_{x'y'\max}\right| = \left|\sigma_{P_1}\right|$, with no normal stress, once again, on these maximum shear planes. For both these states of stress, the Mohr's circle will be centred at the origin. (Similar comments apply for strains.)
- If $\sigma_x = \sigma_y$ along with $\tau_{xy} = 0$, then the state of stress is the same for any rotated x'–y' axes, i.e. $\sigma_{x'} = \sigma_{y'} = \sigma_x$ along with $\tau_{x'y'} = 0$. The corresponding Mohr's circle reduces to a point. (Similar comments apply for strains.)

1.14 Failure Criterion for Brittle Materials

Structures made of brittle materials undergo rupture or crushing when the maximum tensile or compressive stress at a point reaches the ultimate tensile or compressive strength,

$$\text{i.e. } \left|\sigma_{P_1}\right| \text{ or } \left|\sigma_{P_2}\right| \text{ or } \left|\sigma_{P_3}\right| = \sigma_{\text{ult. tensile}} \text{ or } \sigma_{\text{ult. compressive}} \text{ as appropriate.}$$

This is called Maximum Normal Stress Criterion; it is popular because of its simplicity. More accurate criteria exist but are not discussed here.

1.15 Yield Criteria for Ductile Materials

Structures made of ductile materials are designed against yielding; the two commonly used yield criteria are:

(a) **Tresca Criterion**

This states that yielding occurs when the maximum absolute shear stress at a point, as given by

$$\tau_{\text{max. abs.}} = \text{Max}\left[\left|\frac{\sigma_{P_1} - \sigma_{P_2}}{2}\right|, \left|\frac{\sigma_{P_2} - \sigma_{P_3}}{2}\right|, \left|\frac{\sigma_{P_3} - \sigma_{P_1}}{2}\right|\right]$$

reaches $\sigma_{yp}/2$, where σ_{yp} is the yield point stress of the material.

For the case of plane stress, the above equation reduces to

$$\tau_{\text{max. abs.}} = \text{Max}\left[\left|\frac{\sigma_{P_1} - \sigma_{P_2}}{2}\right|, \left|\frac{\sigma_{P_2}}{2}\right|, \left|\frac{\sigma_{P_1}}{2}\right|\right]$$

(b) **von Mises Criterion**

This is based on the consideration of distortional strain energy at a point, and states that yielding occurs when the von Mises equivalent stress given by

$$\sigma_{\text{VM}} = \sqrt{\frac{\left(\sigma_{P_1} - \sigma_{P_2}\right)^2 + \left(\sigma_{P_2} - \sigma_{P_3}\right)^2 + \left(\sigma_{P_3} - \sigma_{P_1}\right)^2}{2}}$$

reaches σ_{yp}.

For plane stress, the above equation reduces to

$$\sigma_{\text{VM}} = \sqrt{\sigma_{P_1}^2 + \sigma_{P_2}^2 - \sigma_{P_1}\sigma_{P_2}}$$

1.16 Statically Determinate and Indeterminate Structures

Structures wherein the support reactions as well as internal member forces or moments can be found out by using equilibrium considerations alone (i.e. net zero force in any direction and net zero moment about any direction, for the whole structure as well as for any free body cut out from it) are called statically determinate; for such structures, displacement analysis, if required, may be carried out subsequently.

Structures which do not fall in the above category are called statically indeterminate; they are further classified as externally statically indeterminate or internally statically indeterminate or both, when equilibrium considerations alone are inadequate to determine the support reactions or internal forces/moments or both, respectively. For such structures, a visualization of the deformed shape and enforcement of one or more displacement compatibility conditions is a prerequisite for force analysis.

1.17 St. Venant's Principle

This may be stated as follows:

"If a system of loads acting on a small portion of the boundary of a body is replaced by a statically equivalent system, then such a replacement causes significant changes in stresses and strains only in the close vicinity of this loaded portion, within a distance equal to the largest dimension of the loaded portion."

Two systems of loads are statically equivalent if they have the same resultant force and moment.

Thus, by virtue of this principle, long structural members loaded only on the end faces may be analysed on the basis of the resultant end force and end moment without accounting for the actual distribution of these loads on the end faces; similarly, support conditions at the ends may be specified in terms of a notional average displacement or slope or twist of the whole cross-section so as to correspond to the actual pointwise restraint conditions over the cross-sectional area and to result in the same net force and moment reactions. The results of such an analysis would then be valid everywhere along the length of the member except for regions close to the ends (see Fig. 1.5).

The above argument may be extended to segments of long members as well; this facilitates consideration of intermediate loads or loads transmitted through junctions where there is a redistribution due to a geometric discontinuity (a sudden change of cross-sectional shape or size).

In regions close to loads, supports, or geometric discontinuities, the state of stress, likely to be three-dimensional with steep gradients, needs to be analysed rigorously, using the theory of elasticity, with a careful consideration of the actual geometry and how the loads are actually applied. St. Venant's principle enables one to avoid such complications during preliminary analysis of the whole structure using simplified theories based on the strength-of-materials approach. It also enables convenient design of end-fixtures and grips for experiments where the focus is on the behaviour of the structure in far-away regions.

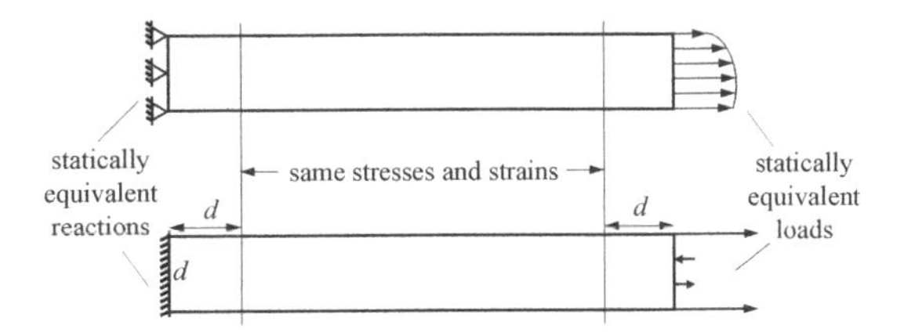

Fig. 1.5 Long thin strip under statically equivalent loads

2 Straight Bars

2.1 Straight Bar

A structural member is classified as a straight bar (Fig. 2.1) when

- it is fairly long as compared to its cross-sectional dimensions and has a straight centroidal axis;
- it is a stand-alone member subjected to loads whose resultants act along the centroidal axis, or part of an assemblage where the member is connected to other members at its ends by pin joints that permit free rotation; in this latter case, it may be a truss member that can resist tension as well as compression, or a wire or string with the loading such as to cause only tension in it;
- it is either stout enough or laterally restrained such that buckling is precluded when subjected to compression.

The bar may be prismatic (i.e. of uniform cross-section along the length), or tapered, or stepped (i.e. made up of several prismatic or tapered segments), and the cross-section may be solid, or hollow tubular or multicellular, or a thin-walled open contour, or any combination of these, and of any arbitrary shape.

K. Bhaskar and T. K. Varadan, *Strength of Materials*,
https://doi.org/10.1007/978-3-031-06377-0_2

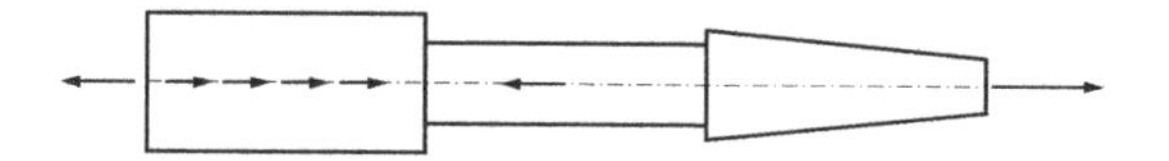

Fig. 2.1 Bar subjected to concentrated and distributed axial loads

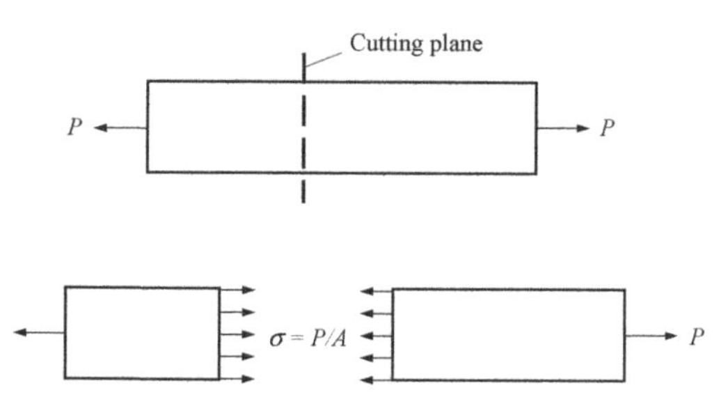

Fig. 2.2 Uniform distribution of axial stress

2.2 Assumptions

- All stresses except the axial stress are negligible.
- The axial stress is uniform across any cross-section (Fig. 2.2).

2.3 Final Formulae

$$\sigma_{\text{axial}} = \frac{P}{A}$$

$$\sigma_{P_1} = \sigma_{\text{axial}};\ \sigma_{P_2} = \sigma_{P_3} = 0;\ \tau_{\text{max. abs.}} = \frac{\sigma_{\text{axial}}}{2}$$

$$\Delta L = \frac{PL}{AE}(\text{for uniform stress})$$

$$\Delta L = \int_0^L \frac{\sigma_{\text{axial}}}{E}\mathrm{d}x(\text{for non-uniform stress})$$

where P and A are the axial force (positive when tensile) and cross-sectional area, respectively, at any section and L is the total length or length of any segment as appropriate. For thin-walled tubular or cellular cross-sections, A is the actual area corresponding to the contour line and not the enclosed area.

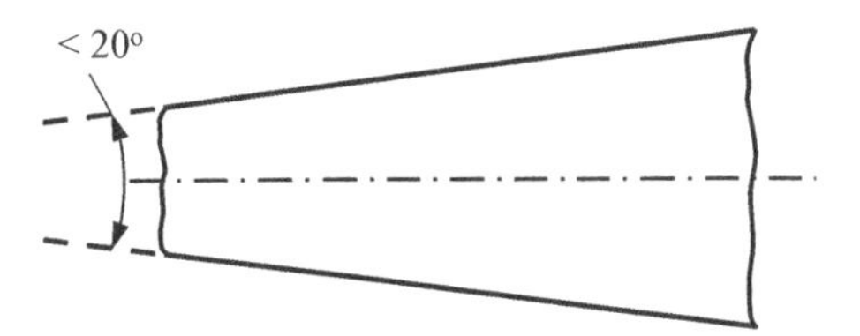

Fig. 2.3 Slight taper

2.4 Range of Applicability

- The formula for the axial stress is not accurate in the vicinity (usually characterized by a distance equal to the largest cross-sectional dimension as per St. Venant's principle) of loaded sections and sudden geometrical changes such as a step or hole; in such regions, the axial stress is not uniform across the cross-section with the maximum stress being much larger than the average value due to stress concentration, and other stress components may also be present.
- The effect of such localized stress variations on calculations of changes in length can be neglected if the bar is fairly long as compared to the largest cross-sectional dimension.
- For tapered bars, the axial stress is not uniform across the cross-section, and other stress components are also present, but such effects may be neglected for taper angles less than 20° as shown in Fig. 2.3.

2.5 Illustrative Problems

Problem 1 Find the stress distribution and displacements of C, D, and G.

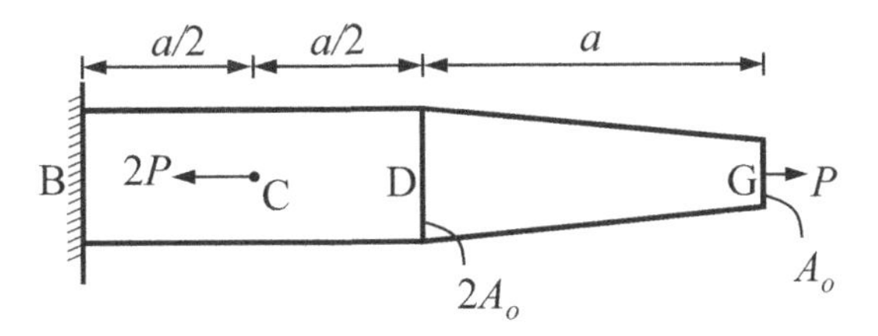

Solution steps:

(1) $\sum F_{\text{axial}} = 0 \Rightarrow$ Reaction at left end $= P$ directed to the right.

(2) Consider free body diagrams as shown.

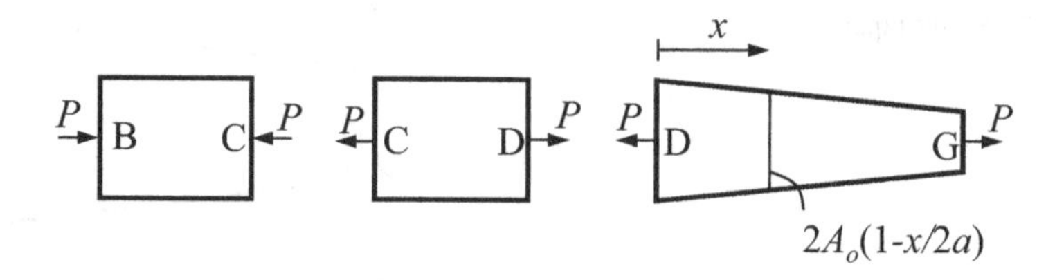

(3) $\sigma_{BC} = -\frac{P}{2A_0};\ \sigma_{CD} = \frac{P}{2A_0};\ \sigma_{DG} = \frac{P}{A(x)} = \frac{P}{2A_0(1-x/2a)}$

(4) $\Delta L_{BC} = -\frac{Pa/2}{2A_0E};\ \Delta L_{CD} = \frac{Pa/2}{2A_0E};$

$$\Delta L_{DG} = \int_0^a \frac{P}{2A_0(1-x/2a)E}dx = \frac{Pa\ln 2}{A_0E}$$

(5) Disp. of $C = \Delta L_{BC} = -\frac{Pa}{4A_0E} = \frac{Pa}{4A_0E}$ to the left.

Disp. of $D = \Delta L_{BC} + \Delta L_{CD} = 0$;

Disp. of $G = \Delta L_{BC} + \Delta L_{CD} + \Delta L_{DG} = \frac{Pa\ln 2}{A_0E}$ to the right.

Problem 2 Find the total elongation of the stepped bar due to self-weight, taking weight density as w.

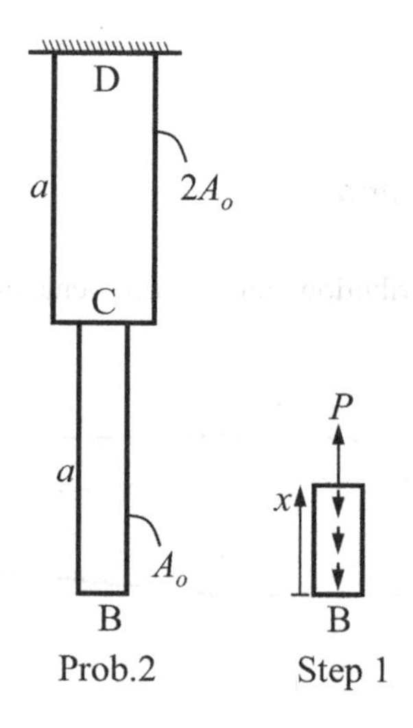

Solution steps:

(1) For a free body of length x ($x < a$) from bottom,

$$P(x) = xA_0w \quad \& \quad \sigma(x) = xw \rightarrow \Delta L_{\mathrm{BC}} = \int_0^a \frac{xw}{E}\mathrm{d}x = \frac{wa^2}{2E}$$

(2) Noting that ΔL_{BC} due to its self-weight is independent of the cross-sectional area, the same result holds good for ΔL_{CD} also due to its self-weight. Besides this, CD undergoes additional elongation due to the weight of BC acting at C, and this is given by $\frac{(aA_0w)a}{2A_0E} = \frac{wa^2}{2E}$

(3) Hence, total $\Delta L = \frac{3wa^2}{2E}$

Problem 3 Find the forces in all the bars and the displacement of the loaded point. All the bars are of the same AE.

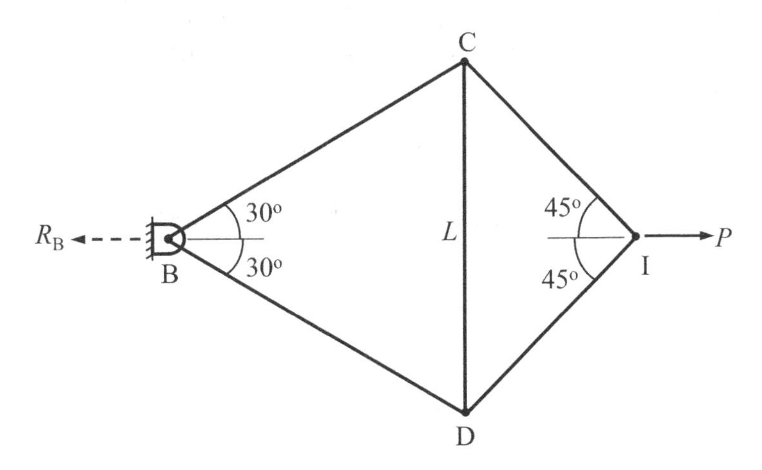

Solution steps:

(1) Global equilibrium yields $R_{\mathrm{B}} = P$
(2) From symmetry considerations, $F_{\mathrm{BD}} = F_{\mathrm{BC}}$, $F_{\mathrm{DI}} = F_{\mathrm{CI}}$.
(3) $\sum H = 0$ at joint I $\Rightarrow 2F_{\mathrm{CI}} \cos 45^\circ = P$

$$\Rightarrow \mathrm{FCI} = P/\sqrt{2}$$

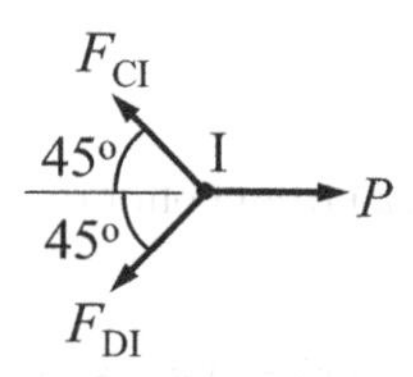

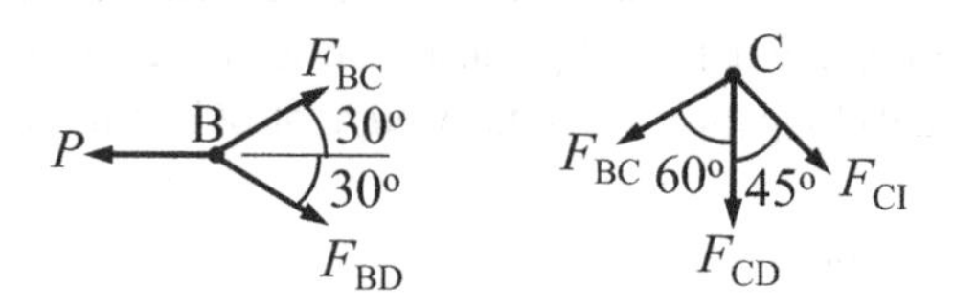

(4) Similarly, $\sum H = 0$ at joint $B \Rightarrow F_{BC} = P/\sqrt{3}$

(5) $\Sigma V = 0$ at joint $C \Rightarrow F_{CD} = -(F_{BC}\cos 60^{o} + F_{CI}\cos 45^{o}) = -\frac{\left(1+\sqrt{3}\right)P}{2\sqrt{3}}$

(6) If G is the mid-point of CD, considering the deformation of right-angled triangle ΔBGC with reference to B, one gets

$$\mathrm{BG}^2 = \mathrm{BC}^2 - \mathrm{CG}^2 \Rightarrow 2\mathrm{BG}\cdot\mathrm{BG} = 2\mathrm{BC}\cdot\mathrm{BC} - 2\mathrm{CG}\cdot\mathrm{CG}$$

i.e.

$$\mathrm{GG}' = \Delta\mathrm{BG} = \frac{\Delta\mathrm{BC}}{\cos 30^\circ} - \Delta\mathrm{CG}\tan 30^\circ = \frac{F_{BC}\mathrm{BC}}{AE\cos 30^\circ} - \frac{F_{CD}\mathrm{CD}\tan 30^\circ}{2AE}$$

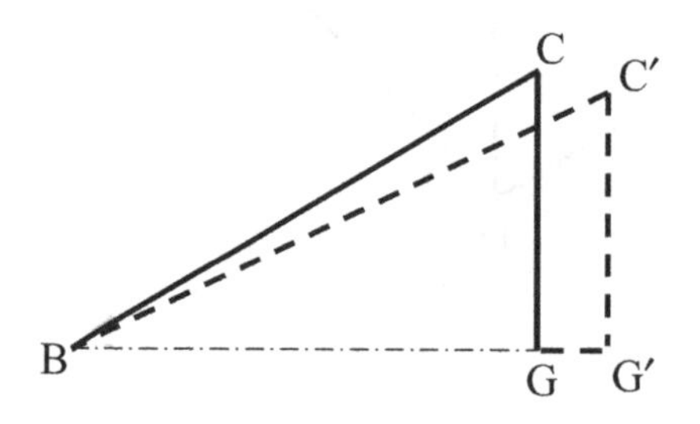

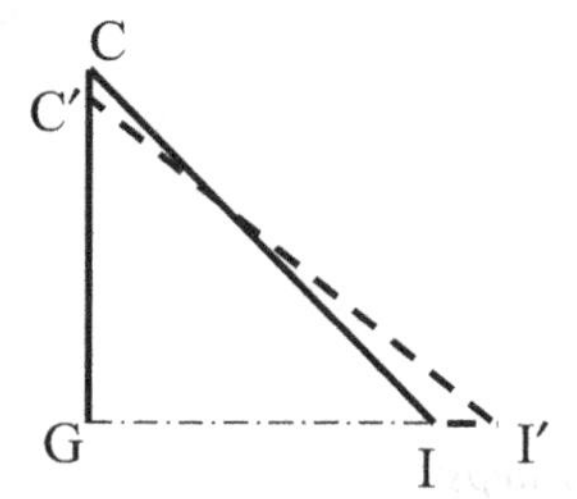

(7) Similarly, considering deformation of ΔCGI with reference to G, one gets

$$\mathrm{II}' = \Delta\mathrm{GI} = \frac{F_{CI}\mathrm{CI}}{AE\cos 45^\circ} - \frac{F_{CD}\mathrm{CD}\tan 45^\circ}{2AE}$$

(8) After due substitutions,

$$\text{displacement of I} = \text{GG}' + \text{II}' = \left(1 + \tfrac{1}{\sqrt{2}} + \tfrac{1}{2\sqrt{3}}\right)\tfrac{PL}{AE} \text{ to the right.}$$

Problem 4 Find the elongation of the compound bar assuming that there is no slip between the central core and the outer shell.

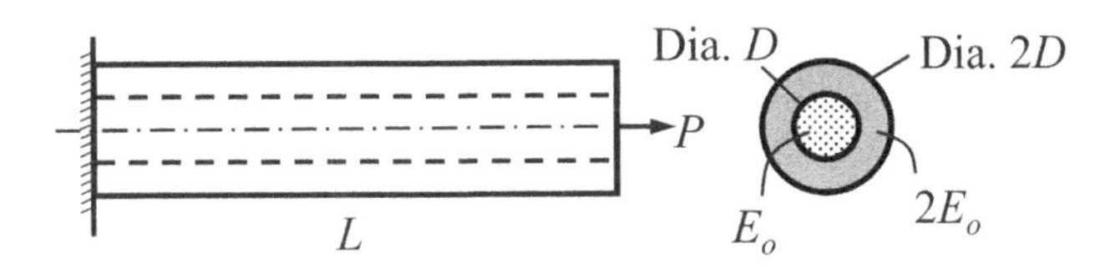

Solution steps:

(1) Net force at any section is P, but its distribution between the core and shell cannot be found out from equilibrium considerations—hence, internally statically indeterminate.
(2) Compatibility condition: $\Delta L_{\text{core}} = \Delta L_{\text{shell}}$,

$$\text{i.e. } \varepsilon_{\text{core}} = \varepsilon_{\text{shell}} = \varepsilon_o(\text{say})$$

(3) $\sigma_{\text{core}} = E_o\varepsilon_o$; $\sigma_{\text{shell}} = 2E_o\varepsilon_o$

(4) Net force $P = \sigma_{\text{core}}A_{\text{core}} + \sigma_{\text{shell}}A_{\text{shell}}$

$$= E_o\varepsilon_o\left(\frac{\pi D^2}{4}\right) + 2E_o\varepsilon_o\left[\frac{\pi(4D^2 - D^2)}{4}\right]$$

Thus, $\varepsilon_o = \frac{4P}{7\pi E_o D^2}$ and $\Delta L = \frac{4PL}{7\pi E_o D^2}$.

Problem 5 Find all the support reactions.

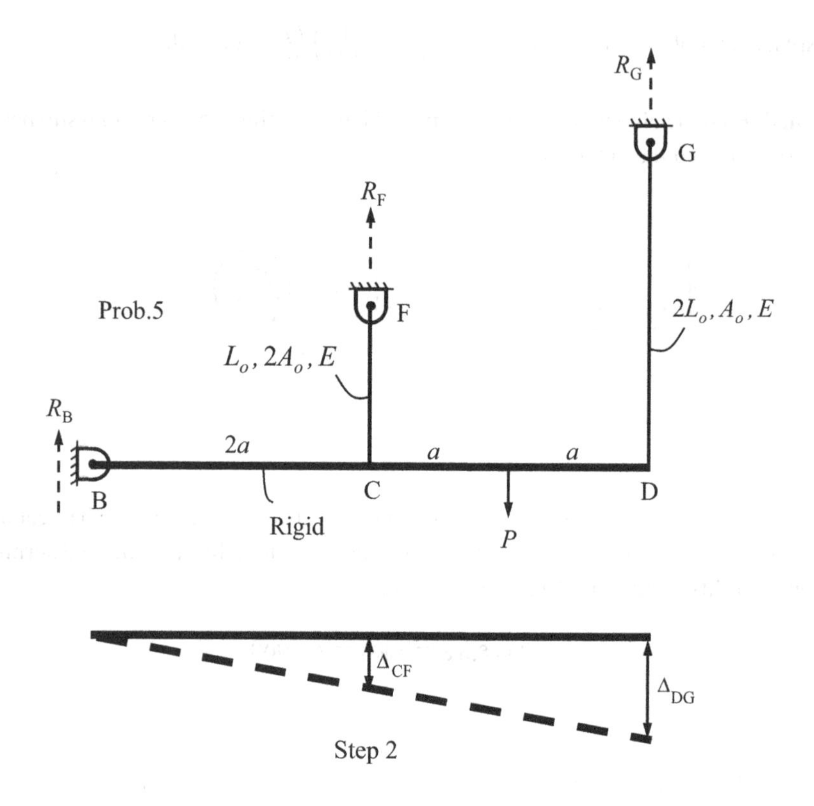

Solution steps:

(1) Reaction at B is vertical since there are no horizontal forces to equilibrate. Three unknown reactions (R_B, R_F, R_G) and two equilibrium equations ($\Sigma V = 0$, $\Sigma M = 0$) $\Rightarrow$ externally statically indeterminate.

(2) Visualization of deformed position $\rightarrow$ compatibility of elongations as given by

$$\Delta_{DG} = 2\Delta_{CF} \Rightarrow \frac{R_G 2L_o}{A_o E} = \frac{2R_F L_o}{2A_o E} \Rightarrow R_F = 2R_G$$

(3) ΣM about $B = 0 \Rightarrow R_F\, 2a + R_G\, 4a = P\, 3a \Rightarrow 2R_F + 4R_G = 3P$
Hence, $R_G = 3P/8$, $R_F = 3P/4$

(4) $\Sigma V = 0 \Rightarrow R_B = P - R_F - R_G = -P/8$, i.e. $P/8$ downwards.

Problem 6 Find the displacement of point D.

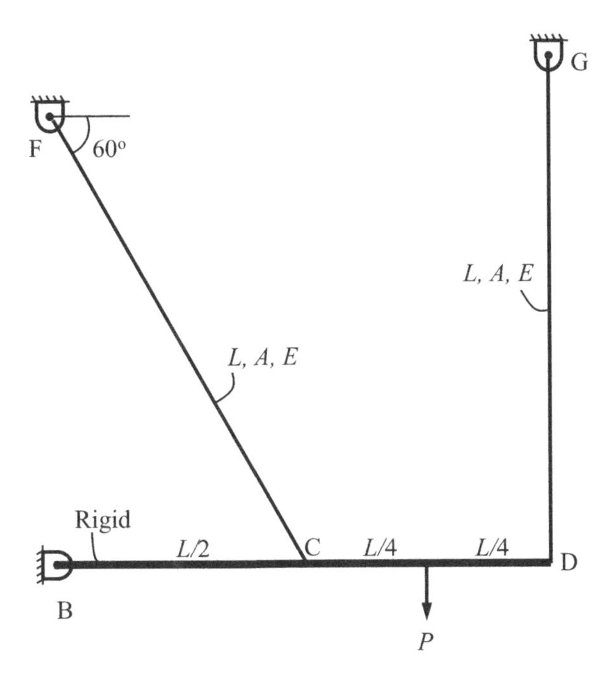

Solution steps:

(1) This is similar to Problem 5 except that member FC rotates about F as it elongates and takes the final position FC_1 as shown; due to rotation alone, C moves to C_2 with the arc CC_2 nearly straight and perpendicular to both FC and FC_1 (because it is small), and due to elongation, it moves from C_2 to C_1.

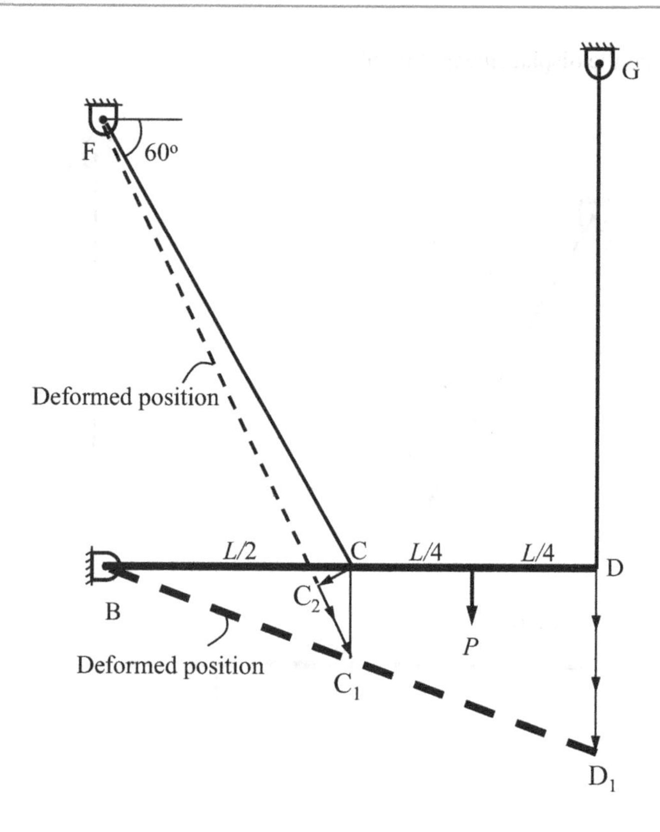

(2) Thus, from ΔCC_1C_2, $\angle C_1CC_2 = 60°$ and $CC_1 = \frac{C_1C_2}{\sin 60°} = \frac{2\Delta_{CF}}{\sqrt{3}}$

(3) $DD_1 = 2(CC_1) \Rightarrow \Delta_{DG} = \frac{4\Delta_{CF}}{\sqrt{3}} \Rightarrow F_{CF} = \frac{\sqrt{3}F_{DG}}{4}$

(4) ΣM about $B = 0 \Rightarrow F_{CF} \sin 60° \left(\frac{L}{2}\right) + F_{DG}L = P\left(\frac{3L}{4}\right)$
yielding $F_{DG} = \frac{12P}{19}$.

(5) Hence, displacement of $D = \Delta_{DG} = \frac{12PL}{19AE}$ vertically downwards.

Problem 7 Find horizontal and vertical displacement components of point C.

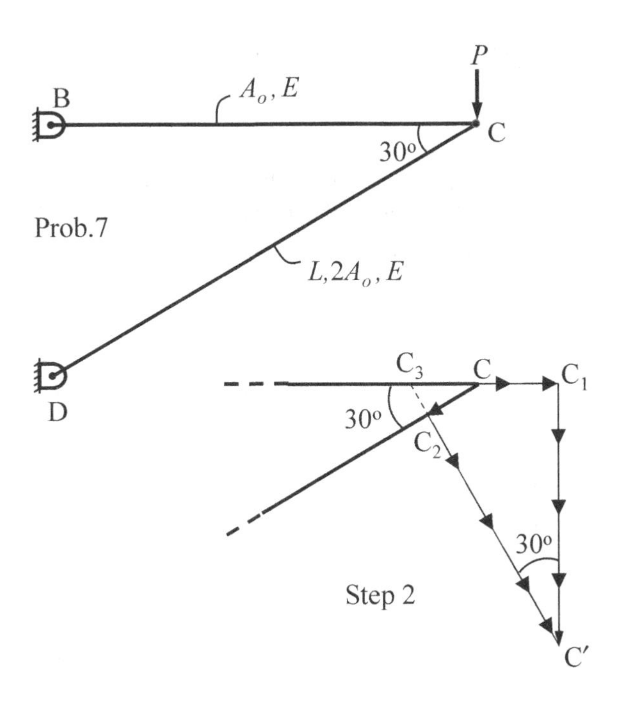

Solution steps:

(1) Enforcing equilibrium of joint C, one gets

$$F_{\mathrm{CD}} = -2P,\ F_{\mathrm{BC}} = P\sqrt{3}$$

(2) As part of BC, C moves to C_1 due to elongation and then to its final position C' due to rotation about B; similarly, as part of CD, C moves to C_2 due to compression and then to C' due to rotation about D (see figure with displacements highly exaggerated).

Thus, horizontal displacement of C is

$$CC_1 = \Delta_{\mathrm{BC}} = \frac{F_{\mathrm{BC}}(L\cos 30^{\circ})}{A_o E} = \frac{3PL}{2A_o E}$$

(3) Extending $C'C_2$ to C_3 as shown, from ΔCC_2C_3,

$$CC_3 = \frac{CC_2}{\cos 30^{\circ}} = \frac{|\Delta_{\mathrm{CD}}|}{\cos 30^{\circ}} = \frac{|F_{\mathrm{CD}}|L}{2A_o E\cos 30^{\circ}} = \frac{2PL}{\sqrt{3}A_o E}$$

(4) From $\Delta C_3C_1C'$, vertical disp. of C is

$$C_1C' = C_3C_1 \cot 30^\circ = (CC_1 + CC_3)\cot 30^\circ = \frac{\left(4+3\sqrt{3}\right)PL}{2A_oE}$$

Problem 8 Given $E_{\text{steel}} = 3E_{\text{Al}}$ and that the coefficients of thermal expansion are related as $\alpha_{\text{Al}} = 2\alpha_{\text{steel}}$, find vertical displacement of the rigid horizontal member due to a temperature increase of ΔT. All the vertical bars have the same cross-sectional area A.

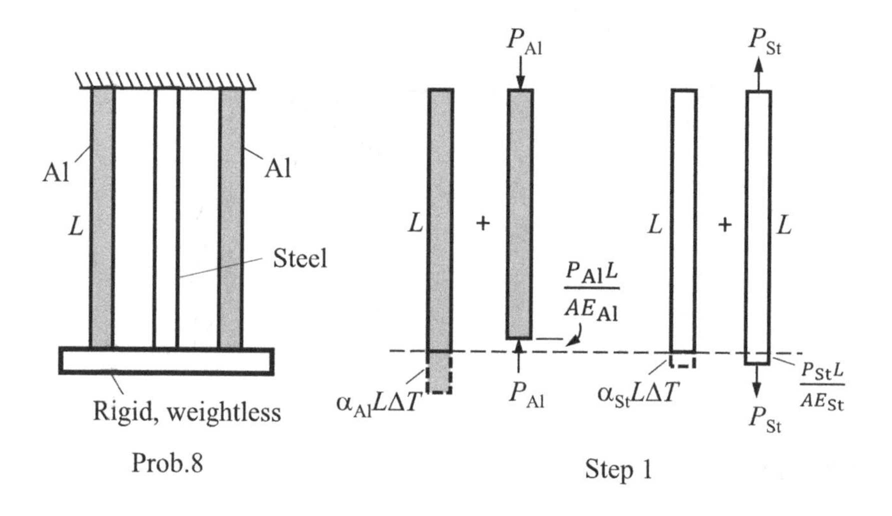

Solution steps:

(1) Net elongation is same for all three vertical bars though free thermal expansion is more for Al than for steel. Hence, a compressive force is generated in Al bars and a tensile force in steel bar as shown.

(2) For equilibrium, $2P_{\text{Al}} = P_{\text{Steel}}$

(3) Thus,

$$\alpha_{\text{Al}}L\Delta T - \frac{P_{\text{Al}}L}{AE_{\text{Al}}} = \left(\frac{\alpha_{\text{Al}}}{2}\right)L\Delta T + \frac{(2P_{\text{Al}})L}{A(3E_{\text{Al}})} \Rightarrow P_{\text{Al}} = \frac{3}{10}AE_{\text{Al}}\alpha_{\text{Al}}\Delta T$$

(4) Required answer = net elongation of any bar

$$= \alpha_{\text{Al}}L\Delta T\left(1 - \frac{3}{10}\right) = \frac{7}{10}\alpha_{\text{Al}}L\Delta T$$

3 Thin-Walled Pressure Vessels

3.1 Thin-Walled Vessels

Attention is confined here to

- cylindrical and spherical vessels of uniform wall thickness, the most commonly encountered cases in practice.
- cases where wall thickness is quite small as compared to the radius.
- cylindrical vessels with length much larger than the diameter.
- cylindrical vessels with closed ends or with open ends (e.g. cylinder with a piston).
- net internal or external pressure loading assuming that buckling does not occur in the latter case.

The vessels are referred to convenient orthogonal r-θ-z axes where r is the radial coordinate, and θ and z are circumferential (or hoop) and axial (or longitudinal) directions for the cylindrical vessel while they are any arbitrary orthogonal tangential directions for the spherical vessel.

3.2 Assumptions

- Radial stress σ_r is negligible.
- The normal stresses σ_θ and σ_z in the wall do not vary through the thickness; because of this assumption, they are referred to as membrane stresses.
- The shear stress τ_{rz} is negligible in the cylindrical vessel.

K. Bhaskar and T. K. Varadan, *Strength of Materials*,
https://doi.org/10.1007/978-3-031-06377-0_3

It should be noted that all the shear stresses are zero in the spherical vessel due to axisymmetry, and $\tau_{r\theta}$ and $\tau_{\theta z}$ are similarly zero in the cylindrical vessel; thus, the shell wall is in a state of plane stress.

3.3 Final Formulae

For cylindrical vessel (Fig. 3.1):

$$\sigma_\theta = \frac{pR}{t}, \sigma_z = \frac{pR}{2t} \text{ (closed ends) or 0 (open ends)}$$

$$\sigma_{P_1} = \sigma_\theta; \quad \sigma_{P_2} = \sigma_z; \quad \sigma_{P_3} \approx 0; \quad \tau_{\text{max. abs.}} = \frac{pR}{2t}$$

$$\left.\begin{aligned} \varepsilon_z &= \tfrac{\Delta L}{L} = \tfrac{pR}{2tE}(1-2\mu) \\ \varepsilon_\theta &= \tfrac{2\pi\Delta R}{2\pi R} = \tfrac{\Delta R}{R} = \tfrac{pR}{2tE}(2-\mu) \end{aligned}\right\} \text{ for closed ends}$$

$$\varepsilon_z = 0, \quad \varepsilon_\theta = \frac{\Delta R}{R} = \frac{pR}{tE} \text{ for open ends}$$

For spherical vessel (Fig. 3.2):

$$\sigma_\theta = \sigma_z = \frac{pR}{2t}$$

$$\left(\sigma_{P_1}, \sigma_{P_2}\right) = (\sigma_\theta, \sigma_z); \quad \sigma_{P_3} \approx 0; \quad \tau_{\text{max. abs.}} = \frac{pR}{4t}$$

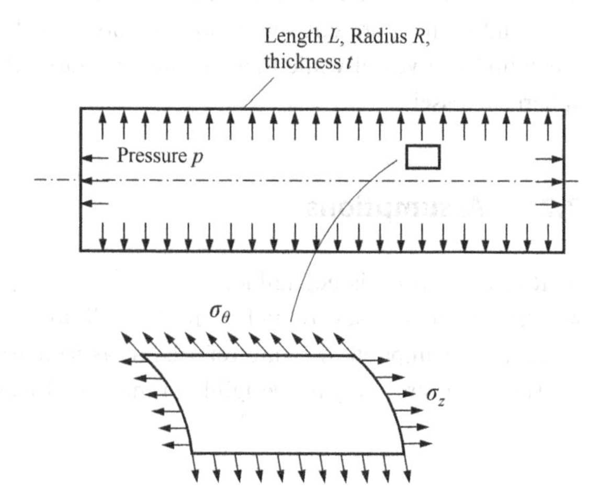

Fig. 3.1 Closed cylindrical vessel

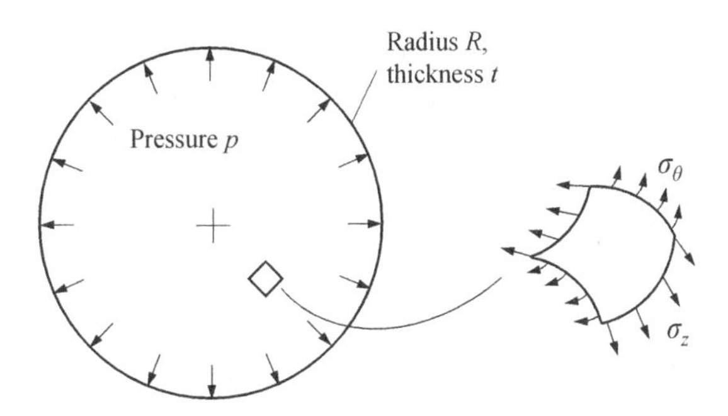

Fig. 3.2 Spherical vessel

$$\varepsilon_\theta = \varepsilon_z = \frac{2\pi \Delta R}{2\pi R} = \frac{\Delta R}{R} = \frac{pR}{2tE}(1-\mu)$$

In the above equations, p is the net outward pressure, R is the radius (inner or mean radius as per convenience), and t is the wall thickness.

3.4 Range of Applicability

- The formulae are acceptable for $R/t > 10$. For thicker vessels, stress variation through the wall thickness can be significant and should be considered.
- The formulae are not applicable in the neighbourhood of geometric discontinuities such as holes or attachments such as lugs. They are also not applicable near the ends of a closed cylindrical vessel where the shape changes suddenly. At such discontinuities and junctions, the stresses may be significantly larger due to localized bending, and the state of stress may be three-dimensional.

3.5 Illustrative Problems

Problem 1: For a closed cylindrical pressure vessel, the strains on the cylindrical wall are measured to be $\varepsilon_z = 100 \times 10^{-6}$, $\varepsilon_\theta = 450 \times 10^{-6}$. Find E and μ if $R = 2$ m, $t = 20$ mm, $p = 0.8$ MPa or megapascals, i.e. 0.8×10^6 Pa.

Solution steps:

(1) $\frac{\varepsilon_\theta}{\varepsilon_z} = \frac{2-\mu}{1-2\mu} = \frac{450}{100} \Rightarrow \mu = 0.3125$
(2) $\varepsilon_\theta = 450 \times 10^{-6} = \frac{pR}{2tE}(2-\mu) \Rightarrow E = 150$ GPa or gigapascals, i.e. 150×10^9 Pa.

Problem 2: A spherical balloon increases in volume by 0.1% from an initial state (defined by radius R_o, internal pressure p_o, and thickness t_o) when the internal pressure is increased by 1%. Assuming this deformation to be linearly elastic, find the corresponding E. Take $\mu = 0.3$.

Solution steps:

(1) $V = \frac{4}{3}\pi R^3 \Rightarrow \frac{\Delta V}{V} = 3\frac{\Delta R}{R}$

(2) Taking the initial state as the reference and the increment in pressure $0.01p_o$ as the load,

$$\frac{\Delta R_o}{R_o} = \frac{0.01 p_o R_o}{2t_o E}(1-\mu) = \frac{0.0035 p_o R_o}{t_o E}$$

(3) Hence, $\frac{3\text{x}\, 0.0035 p_o R_o}{t_o E} = 0.001 \Rightarrow E = \frac{10.5 p_o R_o}{t_o}$

Problem 3: A two-layered thin cylindrical shell is formed by a shrink-fit process where the outer cylinder, of bore slightly smaller than the outer diameter of the inner cylinder, is heated to allow insertion of the inner cylinder and then cooled. If the initial interference of diameters is e, find the stresses due to this shrink fit.

Take the mean diameter of the assembly to be D, the wall thicknesses of the outer and inner cylinders to be equal (t each), and the elastic moduli to be E_o and E_i, respectively.

Solution steps:

(1) Let the pressure between the two cylinders after assembly be p.

(2) Thus, the hoop stresses are:

$$\sigma_{\theta\text{outer}} = -\sigma_{\theta\text{inner}} = \frac{pD}{2t}$$

(3) The corresponding change in diameters should neutralize the initial interference, i.e. $\Delta D_{\text{outer}} + |\Delta D_{\text{inner}}| = e$.

Hence, $\frac{pD^2}{2tE_o} + \frac{pD^2}{2tE_i} = e \Rightarrow \sigma_{\theta\text{outer}} = -\sigma_{\theta\text{inner}} = \frac{pD}{2t} = \frac{eE_oE_i}{D(E_o+E_i)}$.

4 Circular Shafts

4.1 Circular Shaft

This is defined as a structural member (Fig. 4.1)

- of circular or axisymmetric tubular cross-section with the wall thickness large enough to resist buckling;
- of a straight centroidal axis, and one that is prismatic or stepped with fairly long cylindrical or slightly tapered segments as compared to the diameter;
- loaded such that the resultant at any section is a pure twisting moment.

4.2 Assumptions

- Every cross-section behaves as a rigid disc and simply rotates with respect to its undeformed position, or alternatively, radial lines of any cross-section remain straight during twisting, and the cross-section itself remains plane.
- All stresses except the torsional shear stress ($\tau_{z\theta}$ with reference to cylindrical polar coordinates—hereafter denoted by τ without subscripts) are zero.

The implication of these assumptions is that the dimensions of the shaft remain unchanged during twisting and that the torsional shear strain and stress increase linearly along the radius starting from zero at the centroidal axis (Fig. 4.2). It should be noted that there will be a complementary shear stress ($\tau_{\theta z}$) as well at any point.

K. Bhaskar and T. K. Varadan, *Strength of Materials*,
https://doi.org/10.1007/978-3-031-06377-0_4

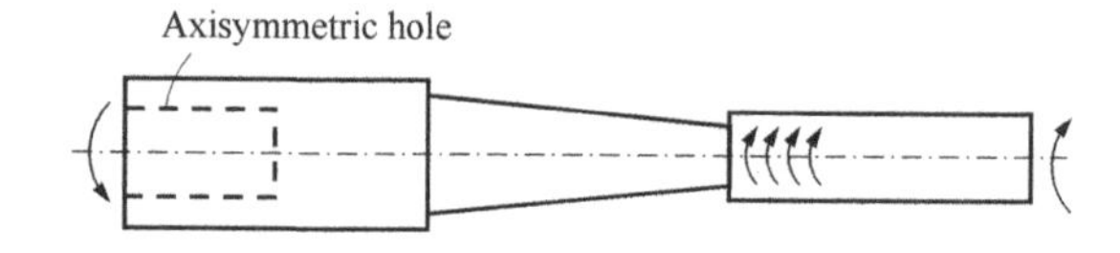

Fig. 4.1 Circular shaft under concentrated and distributed torques

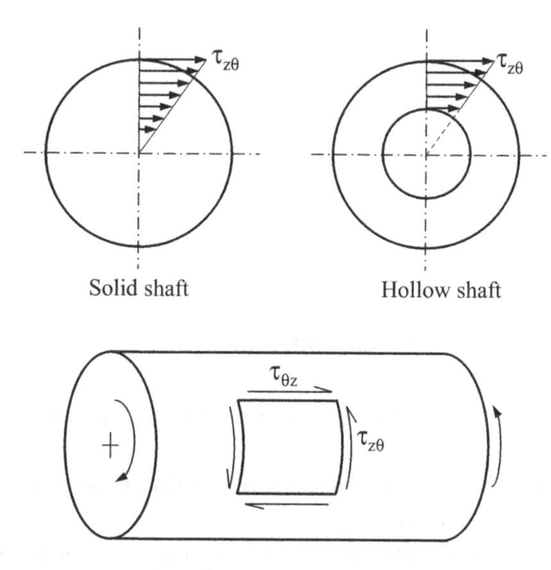

Fig. 4.2 Torsional shear stress

4.3 Final Formulae

$$\frac{\tau}{r} = \frac{T}{J}$$

$$\theta = \frac{TL}{GJ}\text{(for uniform torsion)}, \quad \int_0^L \frac{T}{GJ} dz \text{ (for non - uniform torsion)}$$

where T is the torque at any section and θ is the angle of twist (in radians) of one end with respect to the other end of the shaft (or any of its segments) of length L.

J is the polar moment of inertia of the area of cross-section given by

$$J = \int r^2 \mathrm{d}A = \frac{\pi D^4}{32}\text{(for solid shaft)}, \quad \frac{\pi\left(D_o^4 - D_i^4\right)}{32}\text{(for hollow shaft)}.$$

Thus,

$$\tau_{\max} = \frac{16T}{\pi D^3}\text{or}\frac{16T D_o}{\pi\left(D_o^4 - D_i^4\right)} \quad \text{at } r = \frac{D}{2}\text{or}\frac{D_o}{2} \text{ as appropriate,}$$

and correspondingly,

$$\tau_{\text{max.abs.}} = \sigma_{P_1} = -\sigma_{P_2} = \tau_{\max} \quad \text{as above,} \quad \sigma_{P_3} = \sigma_r = 0$$

For a given $\tau_{\text{allowable}}$, a measure of the torque carrying capacity or torsional strength of a shaft, referred to as polar section modulus Z_p, is defined as

$$Z_p = \frac{T_{\text{allowable}}}{\tau_{\text{allowable}}} = \frac{J}{r_{\max}} = \frac{\pi D^3}{16} \quad \text{or} \quad \frac{\pi\left(D_o^4 - D_i^4\right)}{16 D_o} \quad \text{as appropriate}$$

Similarly, a measure of the resistance of the shaft to twisting deformation, referred to as torsional rigidity, is defined as the torque required to produce unit rate of twist (θ/L or $d\theta/dz$); this is equal to GJ.

If the shaft is used for power transmission, then the transmitted power is given by

$$P = T\omega = \frac{2\pi NT}{60}$$

where ω is the angular speed in rad/sec and N is the corresponding rpm.

The power can be expressed in metric horsepower using the conversion formula.

$$1 \text{ HP} = 75 \text{ kgf m/s} \approx 736 \text{ W}$$

4.4 Range of Applicability

- The shear stress formula is not accurate in the vicinity (usually characterized by a distance equal to the largest cross-sectional dimension as per St. Venant's principle) of loaded sections and sudden geometrical changes such as a step or a circumferential groove; in such regions, the maximum shear stress can be much larger and other stress components may also be present.
- The effect of such localized stress variations on calculations of twist deformation can be neglected if the shaft is fairly long as compared to the largest cross-sectional dimension.
- For tapered shafts or segments in the form of conical frusta, the above formulae are acceptable if the apex angle is less than 10°.

4.5 Illustrative Problems

Problem 1: Find the maximum allowable value of T_o if $G = 80$ GPa and $\tau_{\text{allowable}}$=100 MPa for the material of the shaft, and the rotation of any cross-section should not exceed 3°.

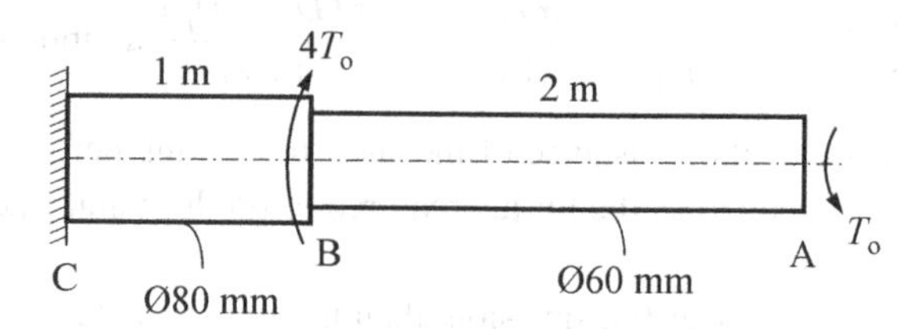

Solution steps:

(1) For the sake of convenience, redesignate the applied torques as T_1 and T_2, *starting from the free end and taking all of them in the same direction,* as shown.

(This step, though not very important for the present problem, will result in error-free calculations when several torques are applied.)

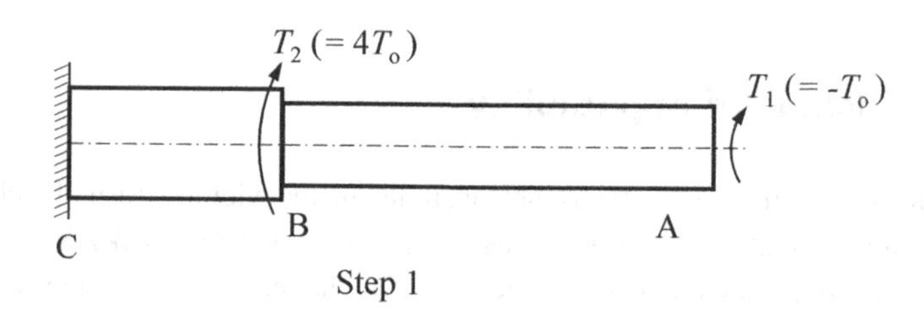

Step 1

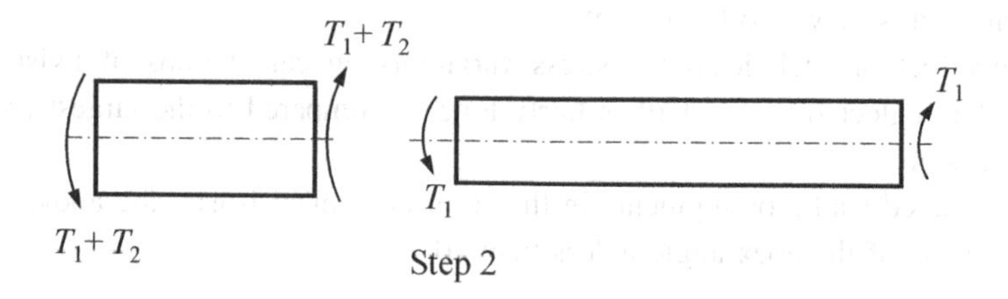

Step 2

(2) Free body diagrams as shown.

(3) $|\tau_{\max}| = \left|\frac{16T}{\pi D^3}\right| = 23579 T_o$ Pa in AB, 29842 T_o Pa in BC, with T_o in N.m. Thus, maximum shear stress occurs in BC, and its value reaches 100 MPa at $T_o = 3351$ N.m.

(4) For each free body, the twist of the right end w.r.t. the left end is calculated as

$$\theta = \frac{TL}{GJ} = -19.65\text{x}10^{-6} T_o \text{for AB, } 9.33\text{x}10^{-6} T_o \text{for BC}$$

where a positive sign indicates clockwise rotation as seen from the right.

Now, *starting from the fixed end*, the rotations of the cross-sections at B and A are calculated as

$$9.33 \times 10^{-6} T_o \quad \text{and} \quad (9.33 - 19.65) \times 10^{-6} T_o = -10.32 \times 10^{-6} T_o$$

Thus, rotation of section A is larger in magnitude, and it reaches $\frac{3\pi}{180}$ rad. at $T_o = 5074$ N m.

(5) Thus, allowable $T_o = \text{Min}(3351, 5074) = 3351$ N.m.

Problem 2: Compare the torsional strength and torsional rigidity of a solid shaft of diameter D with that of a hollow shaft of outer diameter $1.2D$ made of the same material and having the same weight per unit length. Which is better and why?

Solution steps:

(1) For same cross-sectional area,

$$D_o^2 - D_i^2 = (1.2D)^2 - D_i^2 = D^2 \Rightarrow D_i = 0.663D$$

(2) Ratio of torsional strengths is obtained by comparing polar section moduli as

$$\frac{T_{\text{max hollow}}}{T_{\text{max solid}}} = \frac{Z_{p \text{ hollow}}}{Z_{p \text{ solid}}} = \frac{\left(D_o^4 - D_i^4\right)}{D_o D^3} = 1.57$$

(3) Ratio of torsional rigidities is

$$\frac{J_{\text{hollow}}}{J_{\text{solid}}} = \frac{\left(D_o^4 - D_i^4\right)}{D^4} = 1.88$$

(4) The hollow shaft is better because it is more uniformly stressed as compared to the solid shaft wherein the central portion close to the axis is understressed and hence inefficiently utilized.

Problem 3: Find the maximum shear stress in the stepped shaft.

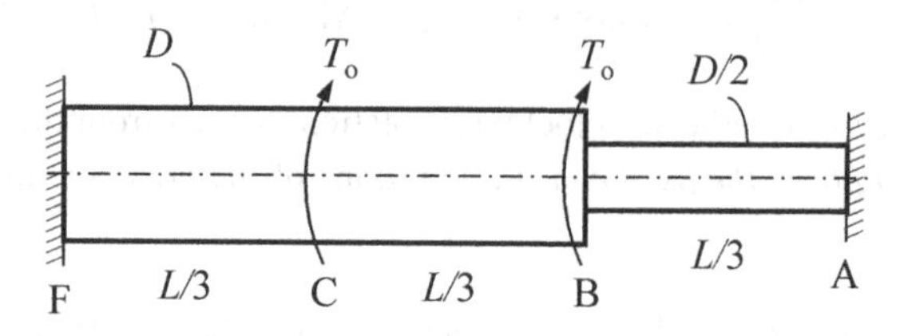

Solution steps:

(1) Two unknown support reactions and one equilibrium equation ⇒ externally statically indeterminate.
(2) Compatibility condition: zero net twist over the entire length.
(3) Denote the right support reaction as T_1 and the applied torques as T_2 and T_3, *all in the same direction,* as shown.

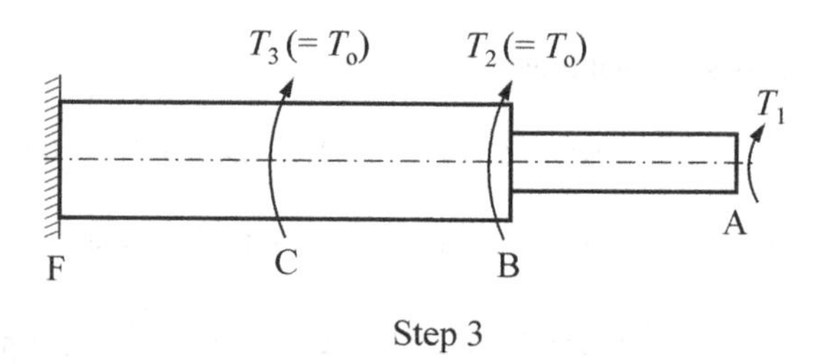

Step 3

(4) Then, the rotation of the cross-section at A is given by

$$\frac{(T_1+T_2+T_3)\left(\frac{L}{3}\right)}{GJ_{\mathrm{CF}}}+\frac{(T_1+T_2)\left(\frac{L}{3}\right)}{GJ_{\mathrm{BC}}}+\frac{T_1\left(\frac{L}{3}\right)}{GJ_{\mathrm{AB}}}$$

with $J_{\mathrm{CF}}=J_{\mathrm{BC}}=\frac{\pi D^4}{32}$, $J_{\mathrm{AB}}=\frac{\pi}{32}\left(\frac{D}{2}\right)^4$.
Equating it to zero yields $T_1=-\frac{T_o}{6}$.

(5) Comparison of $\tau_{\max\,\mathrm{AB}}$, $\tau_{\max\,\mathrm{BC}}$, $\tau_{\max\,\mathrm{CF}}$ yields $|\tau_{\max}|=|\tau_{\max\,\mathrm{CF}}|=\frac{88T_o}{3\pi D^3}$.

Problem 4: Find the maximum power that can be transmitted at 300 rpm through the compound shaft of cross-section as shown assuming that there is no slip between the central core and the outer annulus. Also, plot the corresponding radial variation of shear stress.

Take $\tau_{\mathrm{allowable}}$=75 MPa for steel and 50 MPa for brass, and G=80 GPa and 40 GPa, respectively.

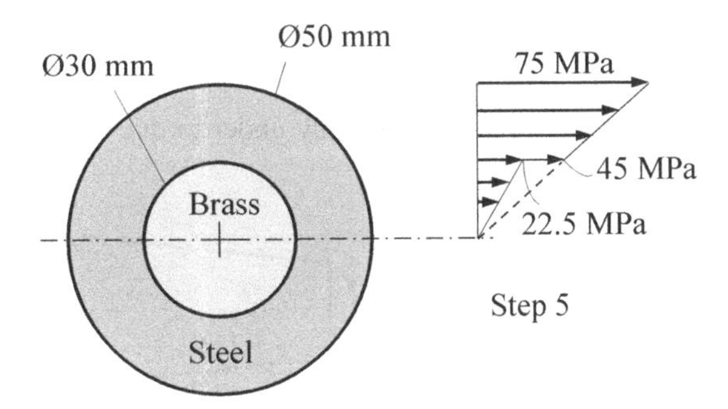

Solution steps:

(1) This is the counterpart of Problem 4 of Chap. 2 on bars and is solved on similar lines.
(2) Compatibility condition: (θ/L) is same for the brass core and the steel annulus.
(3) Hence, $\left(\frac{\tau}{r}\right)_{\text{steel}} = \left(\frac{\tau}{r}\right)_{\text{brass}}\left(\frac{G_{\text{steel}}}{G_{\text{brass}}}\right) = 2\left(\frac{\tau}{r}\right)_{\text{brass}}$
(4) Thus, $\frac{\tau_{\text{max steel}}}{\tau_{\text{max brass}}} = 2\left(\frac{25\text{mm}}{15\text{mm}}\right) = 3.33$
(5) Noting that $\frac{\tau_{\text{all. steel}}}{\tau_{\text{all. brass}}} = 1.5$, it is clear that the shaft can be loaded only till the maximum shear stress in the steel annulus reaches its allowable value of 75 MPa; at this loading, the maximum shear stress in the brass core will be 22.5 MPa which is below the corresponding allowable value. The radial variation of shear stress is as shown.
(6) At this maximum loading,

$$T_{\text{steel}} = \left(\frac{\tau J}{r}\right)_{\text{steel}} = \frac{75}{25}\cdot\frac{\pi\left(50^4 - 30^4\right)}{32}\text{N mm} = 1602.2 \text{ N m}$$

$$T_{\text{brass}} = \left(\frac{\tau J}{r}\right)_{\text{brass}} = \frac{22.5}{15}\cdot\frac{\pi\left(30^4\right)}{32}\text{N mm} = 119.3 \text{ N m}$$

Hence, total T_{max}=1721.5 N m.
Maximum power $= 1721.5\left(\frac{2\pi\times300}{60}\right) = 54083$ W = 73.5 HP.

Problem 5: Find a such that the end reactions are equal.

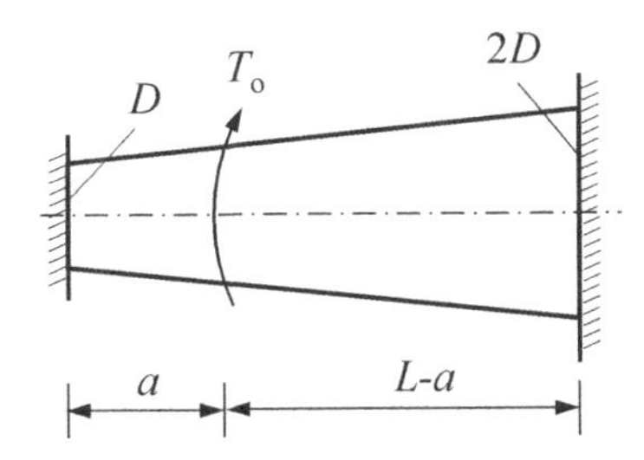

Solution steps:

(1) Free body diagrams as shown with each body under torque $T_o/2$.

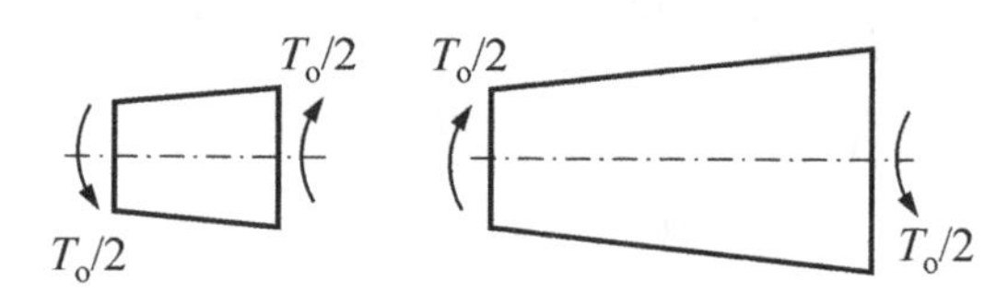

(2) Compatibility condition: total angle of twist is the same for both free bodies.
(3) Diameter at distance x from left end $= D(1 + x/L)$, and hence,

$$J(x) = \frac{\pi D^4}{32}\left(1 + \frac{x}{L}\right)^4$$

(4) $\int_0^a \frac{T_o/2}{GJ(x)}\mathrm{d}x = \int_a^L \frac{T_o/2}{GJ(x)}\mathrm{d}x \Rightarrow \left.\frac{1}{(1+\frac{x}{L})^3}\right|_0^a = \left.\frac{1}{(1+\frac{x}{L})^3}\right|_a^L \Rightarrow a = 0.211L$

Beams—Shear Force and Bending Moment 5

5.1 Beam Undergoing Simple Bending

A beam is defined as a structural member that is fairly long in comparison with the cross-sectional dimensions and subjected to transverse loads. Only beams fulfilling the following conditions are considered in this book:

- The cross-section is at least mono-symmetric.
- The resultant load at any cross-section passes through the plane of symmetry (Fig. 5.1).
- The beam is prismatic or slightly tapered or stepped but with a straight centroidal axis.

Fulfilment of the first two conditions above leads to *simple or symmetrical bending* wherein the deformed centroidal axis of the beam lies in the plane of resultant loading.

In this book, this plane is taken as the x–y plane with x along the longitudinal (horizontal) direction and y along the transverse (vertically downwards) direction with the origin at the centroidal level. The cross-sectional dimensions of the beam are referred to as depth or thickness in the transverse y-direction and breadth or width in the lateral z-direction .

Statically determinate beams - cantilevers or simply supported beams with or without overhangs (Fig. 5.2) — are considered in this chapter. Indeterminate beams will be discussed later (see Sect. 7.6.4).

In general, transverse loads may include point loads and distributed loads in the vertical direction and point moments about the z-direction. Correspondingly, one would get a vertical force as well as a moment about the z-direction as reactions from the fixed end of a cantilever and only vertical force reactions from the supports of a simply supported beam.

K. Bhaskar and T. K. Varadan, *Strength of Materials*,
https://doi.org/10.1007/978-3-031-06377-0_5

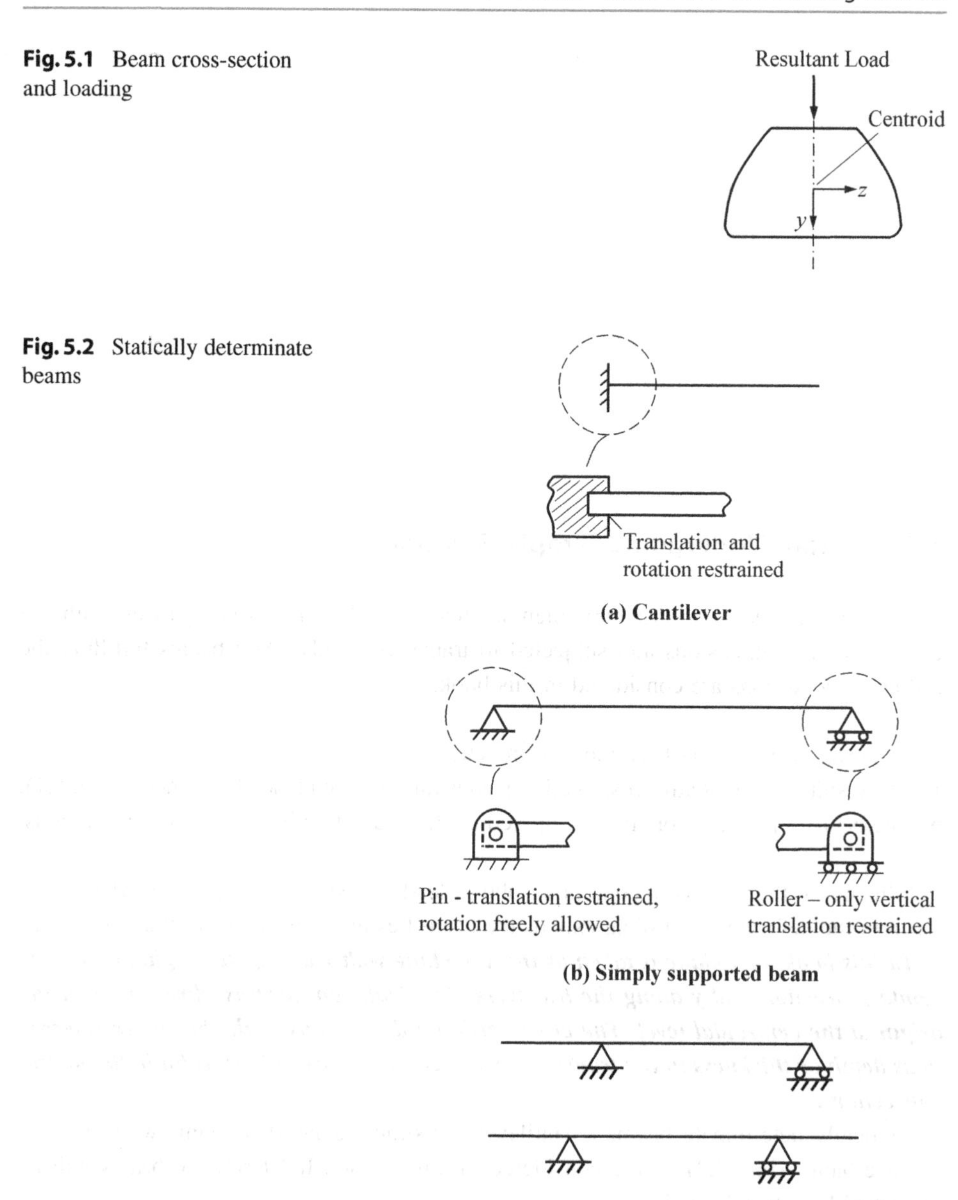

Fig. 5.1 Beam cross-section and loading

Fig. 5.2 Statically determinate beams

5.2 Shear Force and Bending Moment

When a statically determinate beam is cut across a section, one can find out the vertical shear force and the bending moment exerted by one free body on the other through the cut section so that both the free bodies are in equilibrium. This is as shown in Fig. 5.3 for

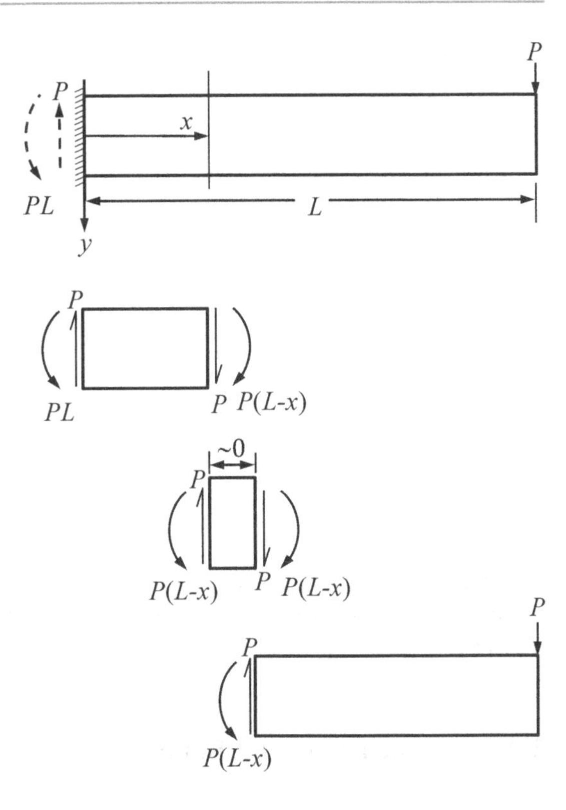

Fig. 5.3 Shear force and bending moment

a particular example; as a general rule, the support reactions are first found out so as to satisfy vertical force and in-plane moment equilibrium for the entire beam, and thereafter, the shear force and bending moment at any section may be calculated by considering either the left free body or the right one.

The internal shear force V and bending moment M may also be shown on an element of infinitesimal length at the chosen distance x from the support as in Fig. 5.3. With respect to such a representation, the sign convention is as indicated in Fig. 5.4; note that positive V corresponds to positive transverse shear stress τ_{xy} across the cross-section, while positive M corresponds to positive or tensile bending stress σ_x at the bottom of the beam (lying on the positive y side) and negative or compressive σ_x at the top of the beam (lying on the negative y side).

5.3 Relations Between Loading, Shear Force, and Bending Moment

$$V = \frac{\mathrm{d}M}{\mathrm{d}x}, \quad q = -\frac{\mathrm{d}V}{\mathrm{d}x}$$

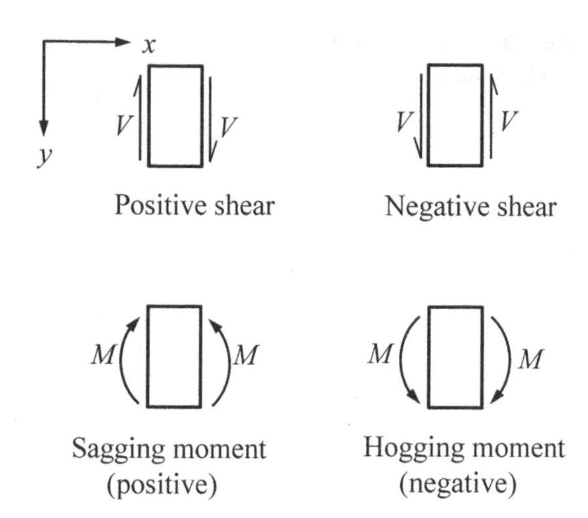

Fig. 5.4 Sign convention

where $q(x)$ is the applied load per unit length, positive when acting downwards (i.e. in the positive y-direction), and $V(x)$ and $M(x)$ are the shear force and bending moment, respectively.

5.4 Shear Force and Bending Moment Diagrams (SFD, BMD)

These diagrams showing variations along the length of the beam are best drawn by visualizing cuts at different sections and finding out the force and moment transmitted by any of the two free bodies through a cut section. The illustrative cases shown in Fig. 5.5 include commonly encountered loads; the support reactions are shown by dotted lines.

While studying these cases, one should note the following:

- On segments with no load/uniform load/linearly varying load, the respective shear force is constant/linearly varying/quadratically varying, and the bending moment variation is linear/quadratic/cubic.
- At the point of application of a concentrated moment, there is a sudden jump in BMD, while SFD is unaffected.
- At the point of application of a concentrated load, there is a sudden jump in SFD and a sudden change of slope in BMD.
- At points corresponding to $V = 0$, slope of BMD is zero.
- The maximum bending moment may occur at points of sudden changes of loading or at a location corresponding to $V = 0$ between such points.

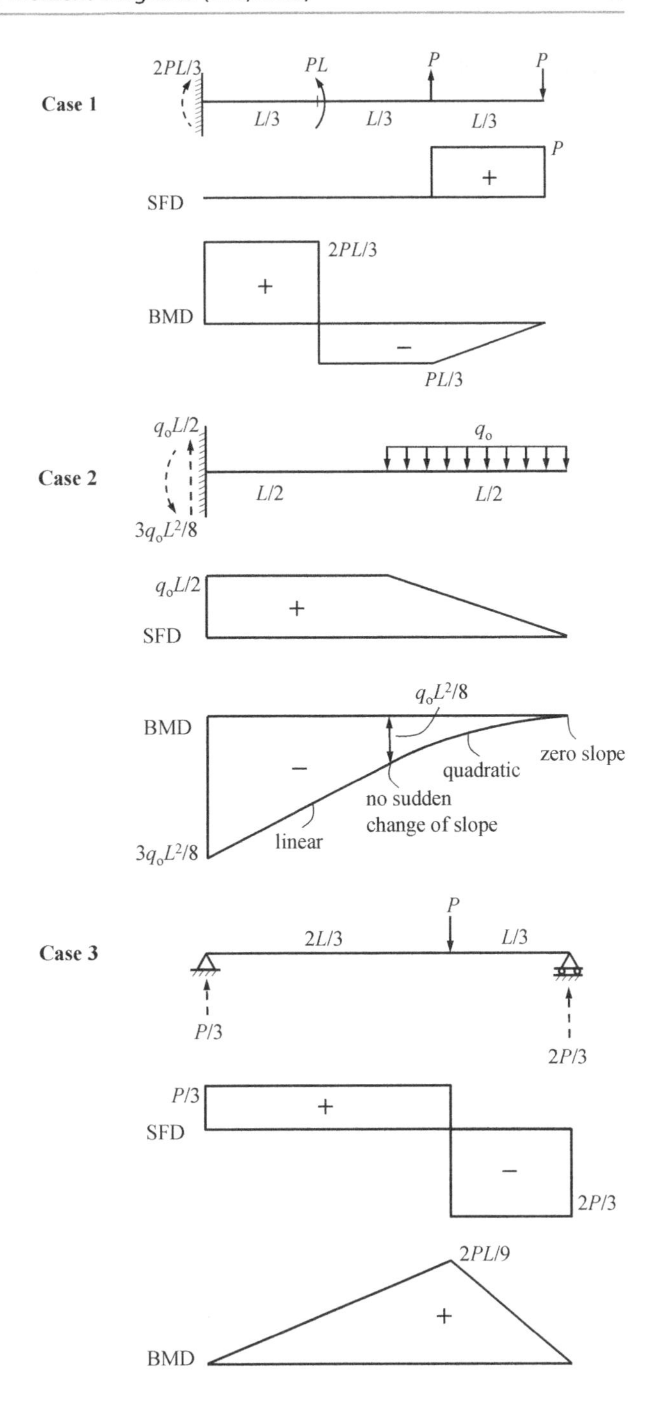

Fig. 5.5 SFD and BMD

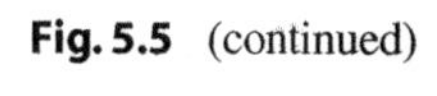
Fig. 5.5 (continued)

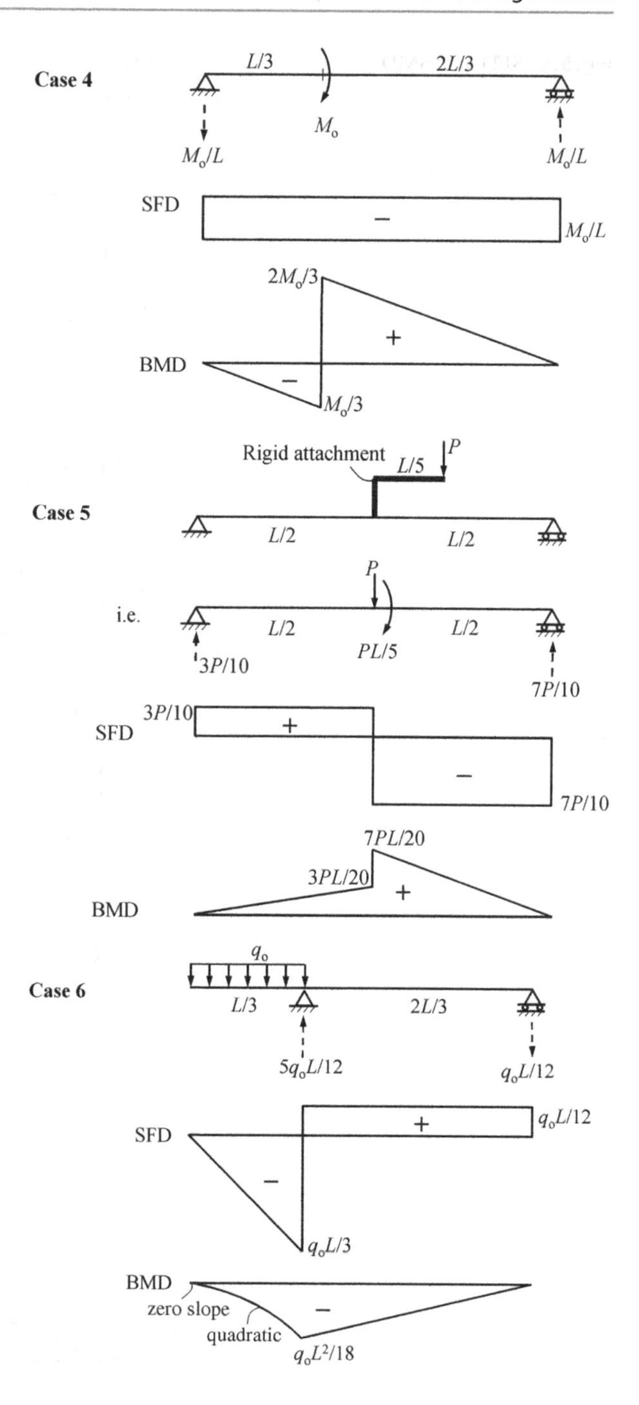

Fig. 5.5 (continued)

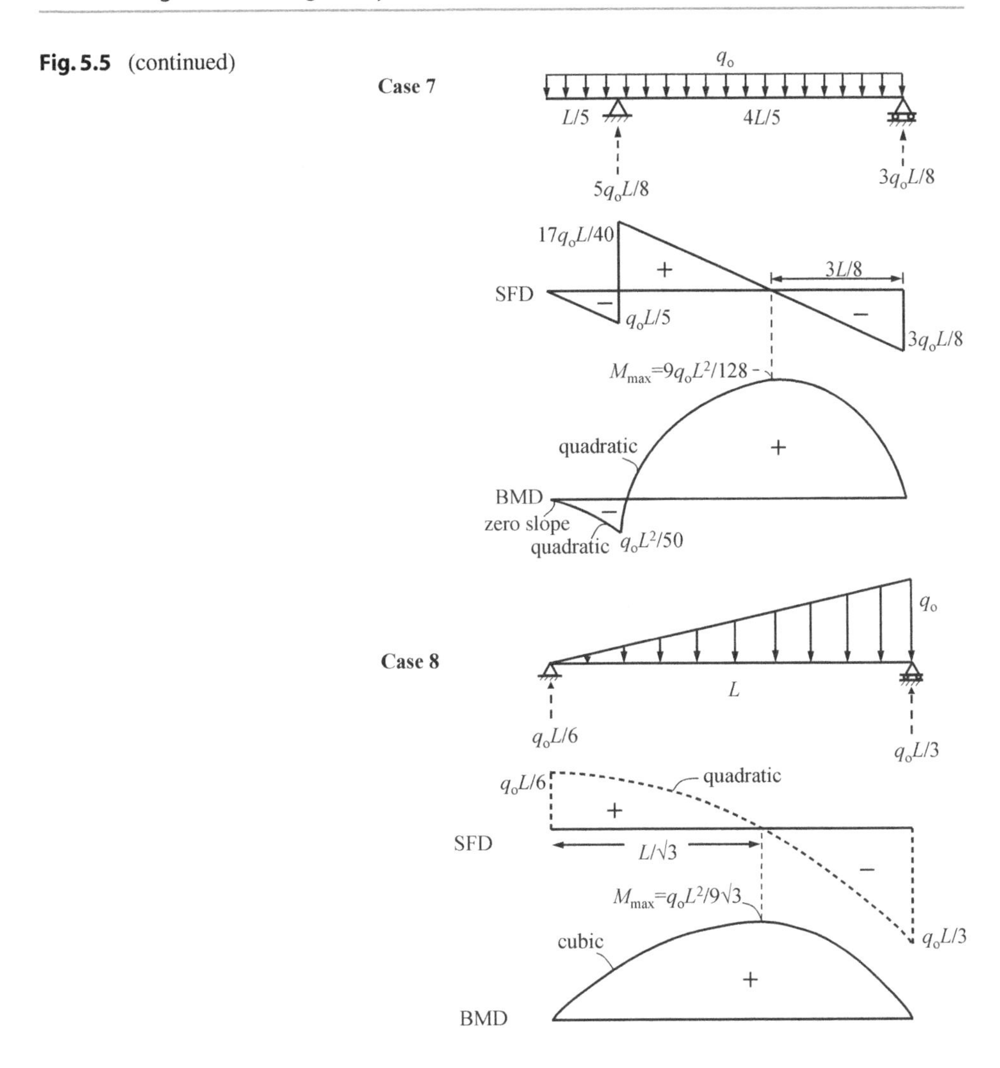

5.5 Bending Moment Diagram by Parts

When there are several loads of different types, it is sometimes convenient (see Sect. 7.6.2) to draw BMD by parts, i.e. corresponding to each load separately starting from the left or right end of the beam. With the sign convention used here, moments due to upward and downward loads or support reactions will be positive and negative, respectively. The net bending moment at any point can be obtained by summing the different contributions, and the net BMD may then be plotted if at all required.

Figure 5.6 shows an illustrative case with BMD drawn by parts starting from the left end as well as from the right end; the equations for the different straight lines or curves

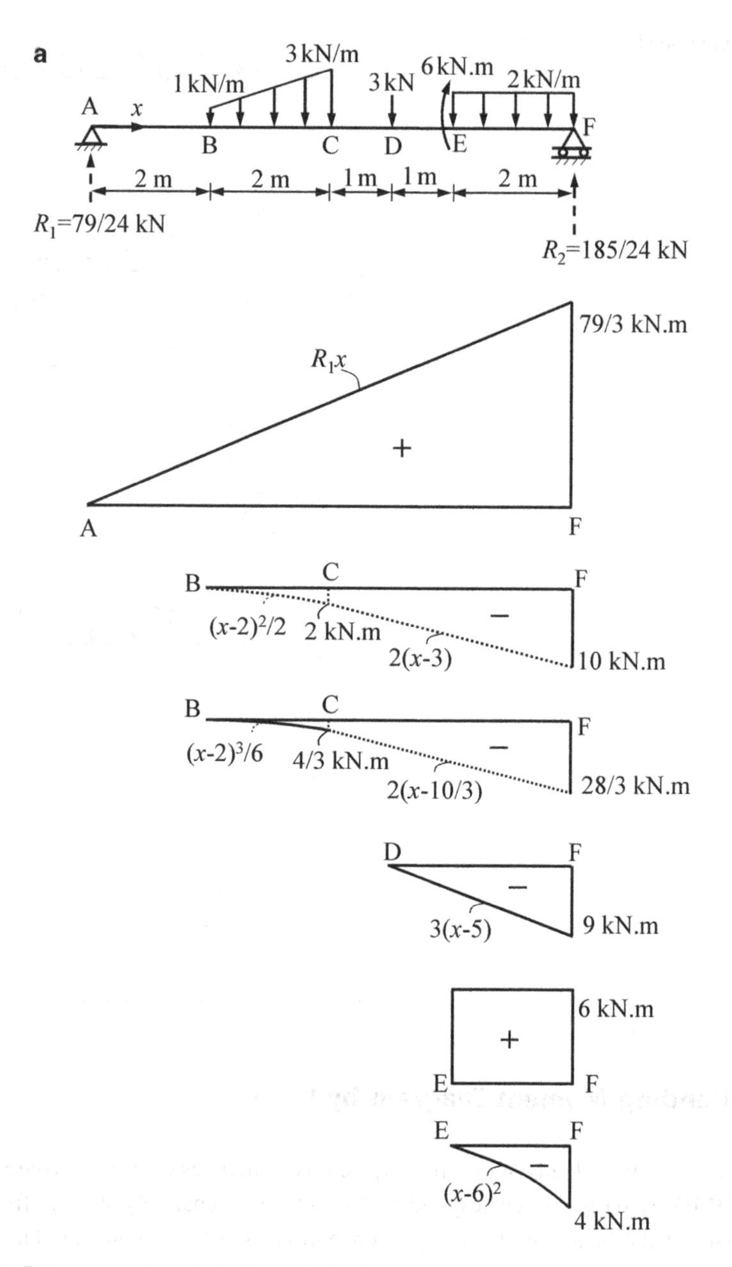

Fig. 5.6 BMD by parts: (**a**) starting from the left end, (**b**) starting from the right end

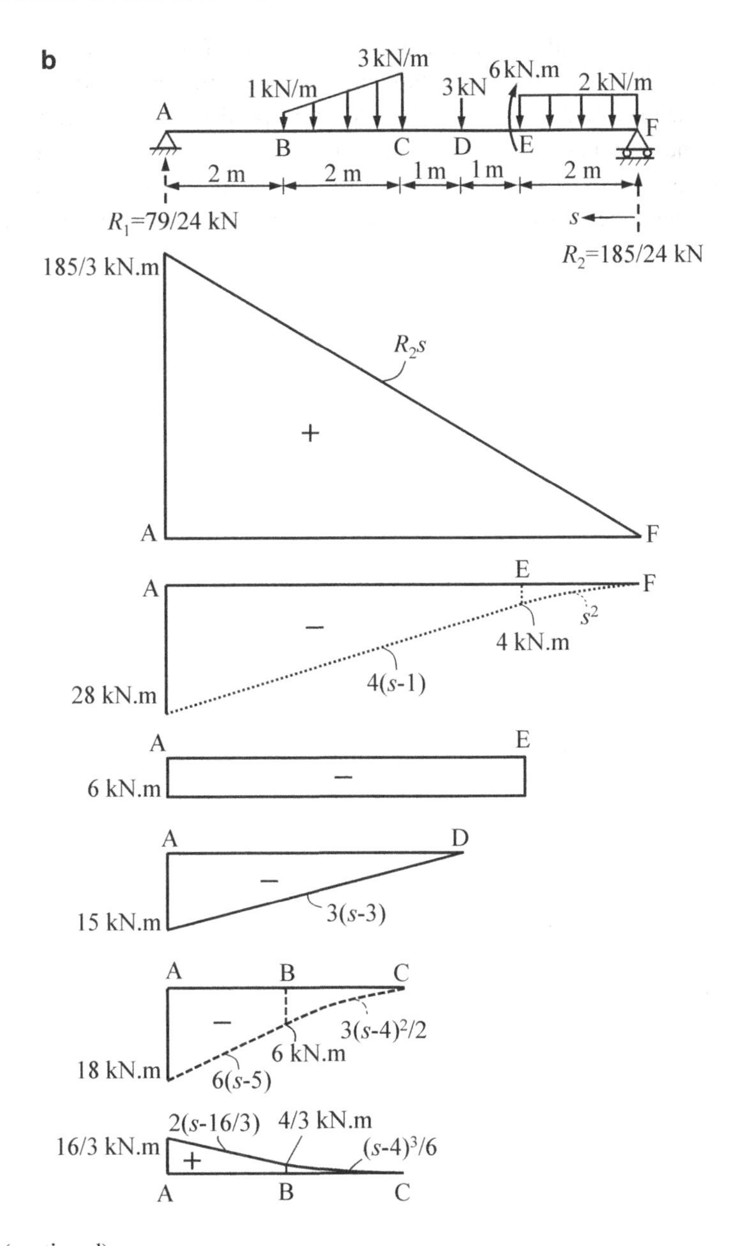

Fig. 5.6 (continued)

are also given. One should note that the trapezoidal load from B to C is conveniently looked upon as either the sum or the difference of a uniform load (1 kN/m or 3 kN/m) and a triangular load (0–2 kN/m or 2 kN/m to 0), respectively. From Fig. 5.6a or b, the values of the net bending moment at points A to F may, respectively, be obtained as 0, 79/12, 59/6, 73/8, (65/12,137/12), 0 kNm. Also, by carefully considering the variation over each segment, the maximum net bending moment may be identified as 137/12 kNm just to the right of E.

6 Beams–Stresses

6.1 Engineering Beam Theory—Assumptions

This theory, presented here for beams described in Sect. 5.1, is based on the following assumptions:

- Plane cross-sections remain plane and perpendicular to the deformed centroidal axis after bending (Fig. 6.1). (This is referred to as Euler–Bernoulli hypothesis. By virtue of this, the transverse shear strain γ_{xy} is zero; however, the transverse shear stress τ_{xy} cannot be zero for transversely loaded beams because it is the stress corresponding to the internal shear force V at any section and thus essential to maintain equilibrium. Thus, a violation of the shear stress–strain constitutive law is explicitly taken for granted by virtue of this assumption.)
- The bending stress σ_x is the dominant normal stress compared to which σ_y and σ_z are negligible. Similarly, shear stress components other than τ_{xy} are negligible. (The two nonzero stress components are often denoted by σ and τ without subscripts.)
- The transverse normal strain ε_y is negligible, i.e. the transverse deflection v does not vary through the depth of the beam. (This is another violation of the constitutive law—between σ_x and its Poisson effect ε_y.)
- Both σ_x and τ_{xy} do not vary across the width of the beam and are hence functions of x and y alone.

6.2 Final Formulae

$$\frac{\sigma_x}{y} = \frac{M}{I}; \quad \tau_{xy} = \frac{VQ}{Ib}$$

K. Bhaskar and T. K. Varadan, *Strength of Materials*,
https://doi.org/10.1007/978-3-031-06377-0_6

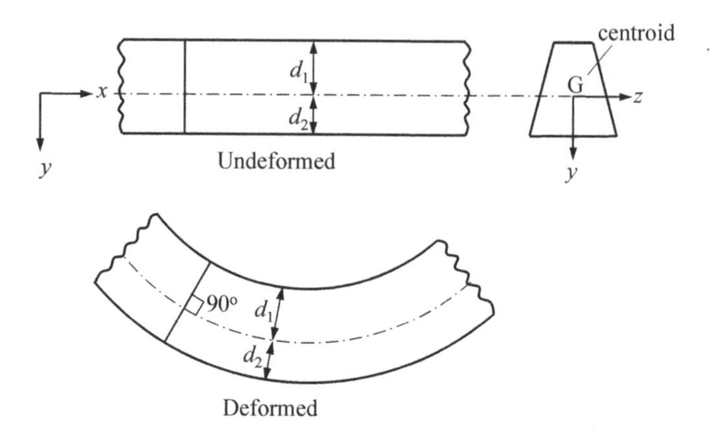

Fig. 6.1 Beam bending as per Euler–Bernoulli theory

$$\varepsilon_x = \frac{\sigma_x}{E}; \quad \varepsilon_z = \frac{-\mu\sigma_x}{E}$$

$$I = \int_{y_{top}}^{y_{bottom}} y^2\,\mathrm{d}A$$

$$Q = \int_{y_{top}}^{y} y\,\mathrm{d}A \quad \text{or} \quad \int_{y}^{y_{bottom}} y\,\mathrm{d}A$$

where I is the moment of inertia of the area of cross-section about the z-axis passing through the centroid (called the *neutral axis* NA—see p. 50), y is the vertical coordinate measured from NA, Q is the moment about NA of the portion of the area of cross-section above or below the level y at which τ_{xy} is evaluated (i.e. the unshaded or shaded area in Fig. 6.2), and b is the width of the cross-section at that level. The neutral axis and the value of I for some common cross-sections are presented in Table 6.1.

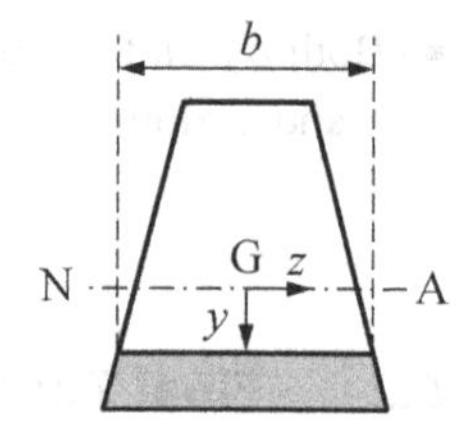

Fig. 6.2 Finding Q

Table 6.1 Moment of inertia I (about NA) for some cross-sections

Rectangle	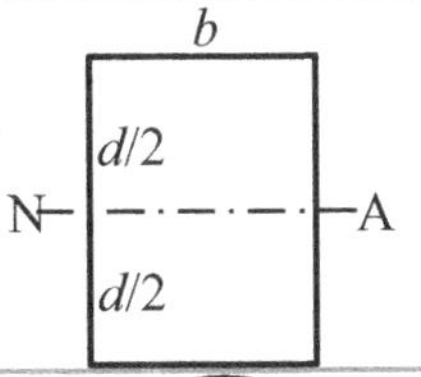 b; d/2; N—A; d/2	$I = \frac{bd^3}{12}$
Circle	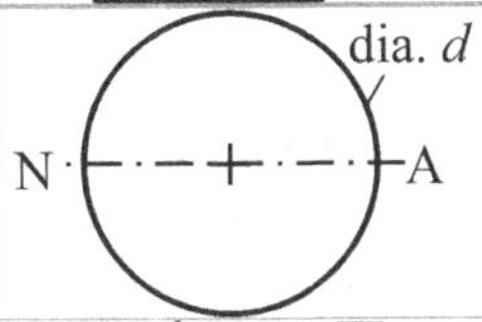 dia. d; N—A	$I = \frac{\pi d^4}{64}$
Isosceles triangle	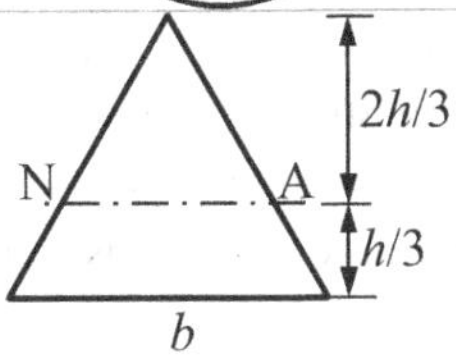 N—A; 2h/3; h/3; b	$I = \frac{bh^3}{36}$
Isosceles trapezoid	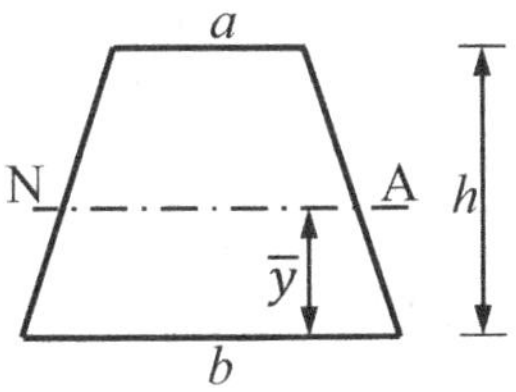 a; N—A; h; $\overline{y}$; b	$\overline{y} = \frac{h(2a+b)}{3(a+b)}$ $I = \frac{h^3\left(a^2+b^2+4ab\right)}{36(a+b)}$
Semi-circle	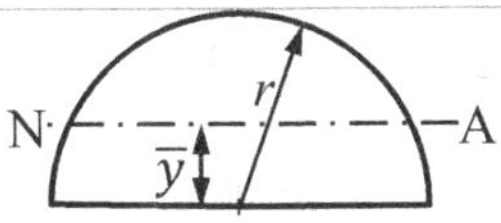 N—A; r; $\overline{y}$	$\overline{y} = \frac{4r}{3\pi}$ $I = \frac{\left(9\pi^2-64\right)r^4}{72\pi}$
Symmetric I-section	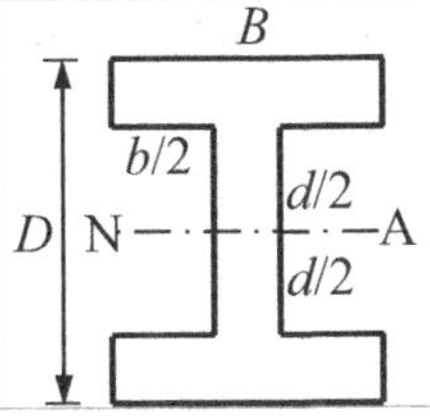 B; b/2; D; N—A; d/2; d/2	$I = \frac{BD^3 - bd^3}{12}$

Except when $b(y)$ is a continuously varying function, Q is easily evaluated as the product of the shaded or unshaded area and the distance of the corresponding centroid from the centroidal axis of the whole cross-section (see Problem 3).

In the above equations, I and Q are always taken to be positive, while σ_x, y, M, and V are algebraic quantities with appropriate signs. Thus, the bending stress distribution

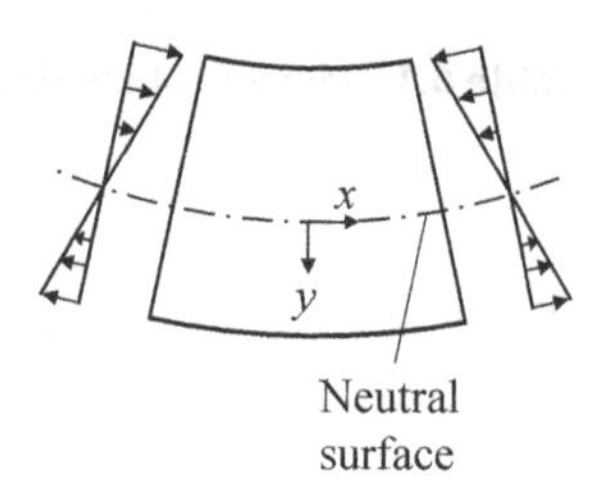

Fig. 6.3 Linear bending stress variation (front view)

for a beam element corresponding to positive or sagging bending moment is as shown in Fig. 6.3. The bending stress is zero at the centroidal level; the corresponding curved surface, between the tensile and compressive zones, is hence strain-free ($\varepsilon_x = \varepsilon_z = 0$) and referred to as the *neutral surface*; its intersection with any cross-section (see Fig. 6.2) is called the *neutral axis* (NA). The bending stress varies linearly with the vertical distance from the neutral axis and thus takes a maximum absolute value for the longitudinal fibres farthest from it; for doubly symmetric cross-sections, the maximum compressive and tensile values are equal in magnitude.

The shear stress formula yields zero values at the top and bottom of the cross-section and a nonlinear variation through the depth of the beam. For a rectangular section of depth d, this variation is parabolic (Fig. 6.4) as given by

$$\tau_{xy} = \tau_{\max}\left(1 - \frac{4y^2}{d^2}\right)$$

For other sections with b varying continuously with y, the variation is more complicated; if b changes suddenly at any level y, the shear stress also undergoes a sudden jump at that level.

For rectangular and solid circular sections, the maximum shear stress occurs at the neutral axis and is given by

$$\tau_{\max} = \frac{3}{2}\tau_{\text{average}} = \frac{3V}{2bd} \quad \text{(for rectangular section)}$$

$$\tau_{\max} = \frac{4}{3}\tau_{\text{average}} = \frac{16V}{3\pi D^2} \quad \text{(for circular section)}$$

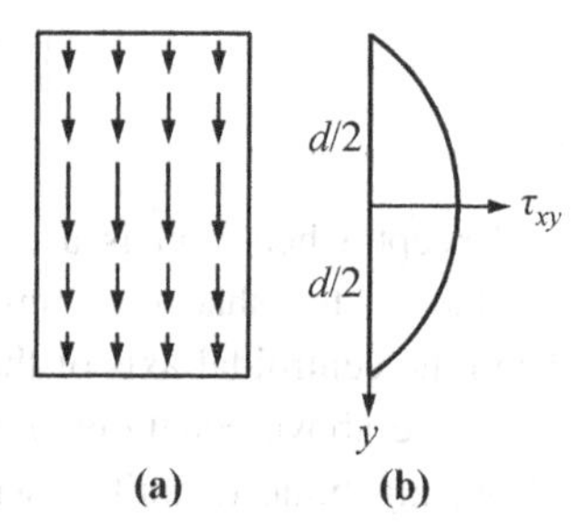

Fig. 6.4 Parabolic shear stress variation in a rectangular section (**a**) Arrows showing direction and magnitude, (**b**) Plot of the magnitude

For other sections, the maximum value may or may not occur at the neutral axis.

For a given $\sigma_{\text{allowable}}$, a measure of the moment carrying capacity or bending strength, referred to as section modulus Z, is defined as

$$Z = \frac{M_{\text{allowable}}}{\sigma_{\text{allowable}}} = \frac{I}{y_{\max}}$$

6.3 Range of Applicability

- Euler–Bernoulli hypothesis is never exactly true for beams subjected to loads causing shear forces, but the associated error is small for length to depth ratio ≥ 20. For shorter beams, shear strain γ_{xy} due to the shear force, and associated warping of the cross-section, should be considered; for very short beams, thickness direction strain (ε_y) should also be accounted for.
- The lateral stress σ_z is negligible for narrow beams but not for wide beams with width (in z-direction) more than 10% of the length. For narrow beams, with negligible σ_z, lateral (width-wise) Poisson effect due to σ_x is freely permitted. This leads to *anticlastic bending* in the lateral direction, whereby a rectangular beam takes the shape of a saddle after deformation (Fig. 6.5). As the cross-section becomes wider, adjacent elements of it restrain each other from such anticlastic bending, and thus, lateral stresses σ_z get generated within the beam.
- The formulae for the bending and shear stresses are not accurate in the vicinity (usually characterized by a distance equal to the largest cross-sectional dimension as per St. Venant's principle) of loaded sections and sudden geometrical changes such as a step, where severe stress concentration may occur along with the presence of other stress components.
- The shear stress formula is not applicable to non-rectangular cross-sections whose lateral faces are not vertical; thus, it should not be applied to triangular, semi-circular, and other such cross-sections. However, for circular cross-sections, it can be used

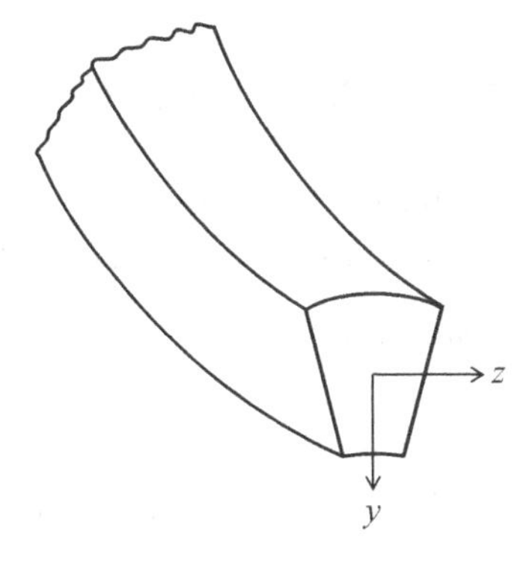

Fig. 6.5 Anticlastic bending of a narrow rectangular beam

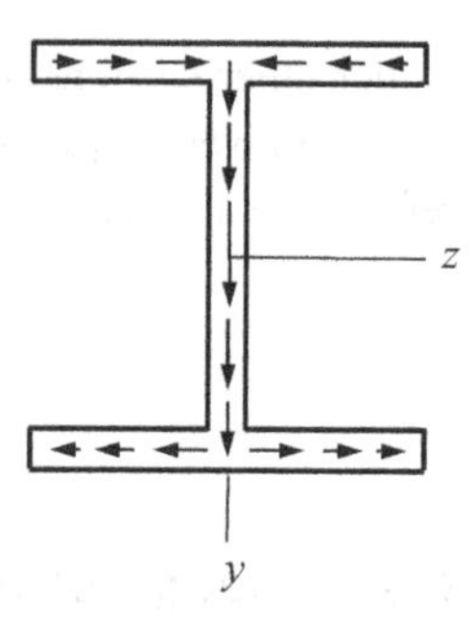

Fig. 6.6 Arrows showing shear stress direction and magnitude in thin-walled I-section subjected to downward shear force

to predict the maximum shear stress which occurs at the level of the horizontal diameter with acceptable accuracy; this is true for hollow circular sections also with $D_{\text{outer}}/D_{\text{inner}} \leq 1.1$.

- Even for rectangular sections, the shear stress formula gives accurate results only if the width b is smaller than half of the depth d; otherwise, the variation of this stress across the width of the section has to be accounted for.
- The shear stress formula should not be used for tapered beams or segments. However, the bending stress formula is reasonably accurate for taper angles (defined as shown in Fig. 2.3) less than 20°.
- For thin-walled open or closed sections, while the bending stress formula given above is applicable without any change, such is not the case with the shear stress formula. This is because the shear stress for such sections is predominantly oriented along the direction of the thin-walled contour; for example, the shear stress orientation in a very thin-walled I-section is as shown in Fig. 6.6, where the horizontal flanges are subjected to significantly high τ_{xz} and negligible τ_{xy}. Similarly, thin-walled sections with inclined segments will be subjected to shear stresses which can be resolved into τ_{xz} and τ_{xy} components with neither of them being negligible. In all such cases, the present shear stress formula which yields τ_{xy} alone is not adequate; however, it can still be used to correctly estimate the shear stresses in vertical segments of such cross-sections.

6.4 Illustrative Problems

Problem 1 Compare the maximum bending and shear stresses in a cantilever of dimensions $L \times b \times d$ subjected to uniform load q_o per unit length, and draw appropriate conclusions.

Solution steps:

(1) $M_{\max} = -\frac{q_o L^2}{2}$, $V_{\max} = q_o L$, both occurring at the fixed end.

(2) So, $|\sigma_{\max}| = \frac{3q_o}{b}\left(\frac{L}{d}\right)^2$, $\tau_{\max} = \frac{3q_o}{2b}\left(\frac{L}{d}\right)$, and, hence, $\frac{\sigma_{\max}}{\tau_{\max}} \propto \frac{L}{d}$.

(3) Thus, for long beams with $(L/d) > 20$, τ_{max} is much smaller than σ_{max}; this is in general true for beams of solid cross-sections under any loading. For ductile materials, the shear stress at the onset of yielding is about 0.5–0.6 times σ_{yp}, and thus, long beam failures are always due to excessive bending stresses. (However, if the beam cross-section is built up by bonding several horizontal layers, then the adhesive between the layers may undergo shear failure due to τ_{max} much before the extreme layers fail due to excessive bending stresses.)

Problem 2 For the beam of inverted *T* cross-section as shown, find the maximum tensile stress if the maximum compressive stress is of magnitude σ_o occurring at the bottom fibre. Find also the corresponding bending moment.

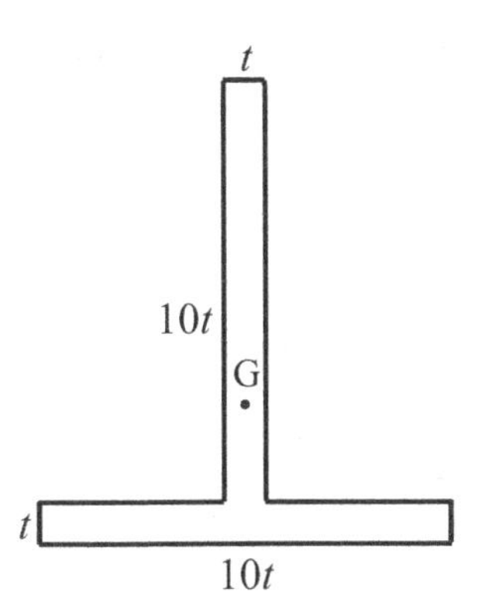

Solution steps:

(1) The bending stress at any level is proportional to its distance from the centroidal axis. So, the centroid *G* of the cross-section has to be located. Its location with respect to the bottom fibre is given by

$$\frac{t \times 10t \times (t + 5t) + 10t \times t \times t/2}{10t \times t + 10t \times t} = 3.25t$$

(2) Maximum tensile stress occurs at the top fibre and is related to that at the bottom fibre as

$$\left|\frac{\sigma_{top}}{\sigma_{bot.}}\right| = \left|\frac{y_{top}}{y_{bot.}}\right| = \frac{11t - 3.25t}{3.25t}$$

and thus $\sigma_{max.\ tensile} = 2.385\sigma_o$

(3) To find M, I is required, and it may be calculated by using parallel axis theorem for each of the constituent rectangles as

$$\frac{t \times (10t)^3}{12} + t \times 10t(11t - 3.25t - 5t)^2 + \frac{10t \times t^3}{12} + 10t \times t\left(3.25t - \frac{t}{2}\right)^2 = 235.42t^4$$

(4) With proper signs,

$$\sigma_{\text{bot.}} = -\sigma_o,\ y_{\text{bot.}} = 3.25t,\ \text{and hence},\ M = 72.44\,\sigma_o t^3$$

Problem 3 For the beam of Problem 2, plot the shear stress variation across the cross-section corresponding to a shear force V showing values at the neutral axis and the web-flange junction.

Solution steps:

(1) At the neutral axis, considering the area above it,

$$Q = \frac{t(11t - 3.25t)^2}{2} = 30.0t^3$$

Thus, $\tau_{\text{NA}} = \frac{VQ}{It} = 0.128\frac{V}{t^2}$

(2) At the web-flange junction, considering the flange area,

$$Q = 10t \times t\left(3.25t - \frac{t}{2}\right) = 27.5t^3$$

Using $b = t$ in web and $10t$ in flange, one gets $\tau = 0.117\frac{V}{t^2}, 0.0117\frac{V}{t^2}$, respectively.

(3) The variation is quadratic in both the web and the flange and is as shown.

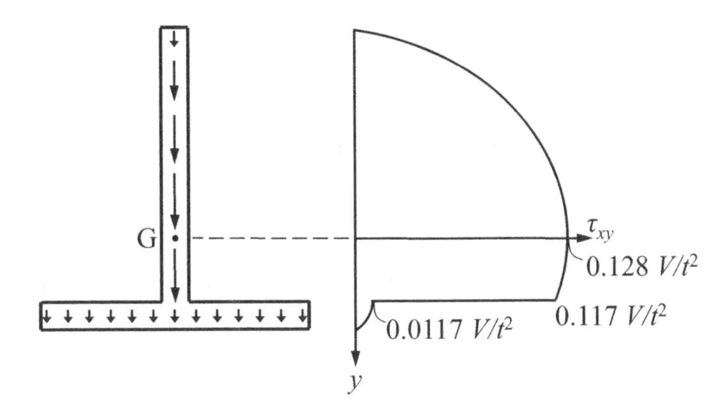

Problem 4 A beam of tubular cross-section as shown is bent vertically by a bending moment M. Find the extreme fibre stresses. Assume the cross-section to be thin-walled, i.e. ignore the spread of the material in the thickness direction and take all of it to be located on the centre line of the cross-sectional contour. (This assumption results in little error in the final results if the thickness is very small compared to the length of the contour segments.)

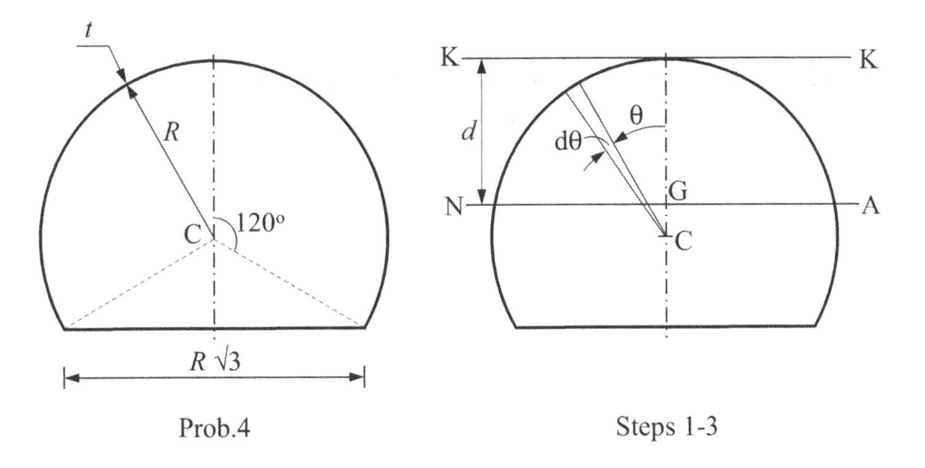

Prob.4 Steps 1-3

Solution steps:

(1) The location of the centroid G is found out by calculating the first moment of the area about the topmost point and dividing it by the total area; for the circular arc, this is done by considering a small element of area $tRd\theta$ at an angular distance of θ from the topmost point as shown. Thus, one gets

$$d = \frac{2\int_0^{\frac{2\pi}{3}} tR.R(1-\cos\theta)\mathrm{d}\theta + tR\sqrt{3}.R\left(1-\cos\frac{2\pi}{3}\right)}{\frac{2}{3}.2\pi Rt + tR\sqrt{3}} = 0.8537R$$

(2) Similarly, the moment of inertia about axis KK through the topmost point is calculated as

$$I_{\mathrm{KK}} = 2\int_0^{\frac{2\pi}{3}} tR.[R(1-\cos\theta)]^2\mathrm{d}\theta + tR\sqrt{3}.\left[R\left(1-\cos\frac{2\pi}{3}\right)\right]^2$$
$$= 2\pi tR^3$$

(3) Hence, $I_{\mathrm{NA}} = I_{\mathrm{KK}} - Ad^2$

$$= 2\pi tR^3 - \left(\frac{2}{3}.2\pi Rt + tR\sqrt{3}\right)(0.8537R)^2 = 1.968tR^3$$

(4) Using $|y_{\mathrm{top}}| = 0.8537R$, $|y_{\mathrm{bottom}}| = 1.5R - 0.8537R$, one gets

$$|\sigma_{\mathrm{top}}, \sigma_{\mathrm{bot.}}| = \left|\frac{My}{I}\right| = (0.434, 0.328)\frac{M}{tR^2}$$

Problem 5 For the beam shown, find the maximum tensile and compressive bending stresses. Take the cross-section as in Problem 2.

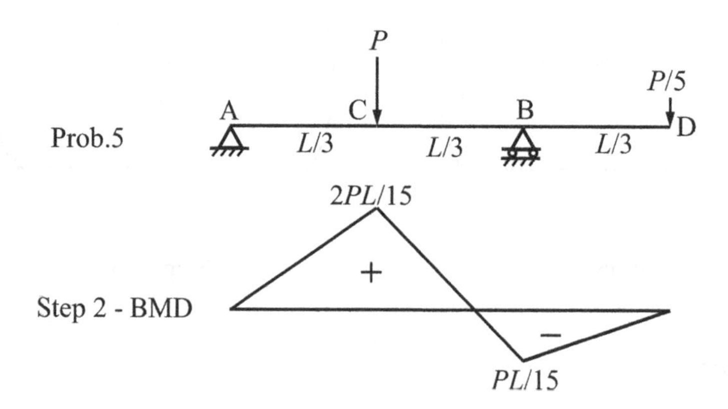

Solution steps:

(1) The support reactions are calculated as $R_{\mathrm{A}} = 2P/5$ and $R_{\mathrm{B}} = 4P/5$, both upwards.
(2) The bending moment diagram as shown.

(3) At section C,

$$\left(\sigma_{\text{top}}, \sigma_{\text{bot.}}\right) = \left(\frac{2PL}{15}\right)\left(\frac{-7.75t, 3.25t}{235.42t^4}\right) = (-4.39, 1.84) \times 10^{-3}\, PL/t^3$$

(4) Similarly, at section B, with $M_B = -PL/15$,

$$\left(\sigma_{\text{top}}, \sigma_{\text{bot.}}\right) = (2.19, -0.92) \times 10^{-3}\, PL/t^3$$

(5) Thus, maximum $(\sigma_{\text{tensile}}, \sigma_{\text{comp.}}) = (2.19, -4.39) \times 10^{-3}\, PL/t^3$, both occurring at the topmost fibre, and at sections B and C, respectively. Though $M_B < M_C$, the maximum tensile stress occurs at section B due to the large difference between y_{top} and $y_{\text{bot.}}$

Problem 6 Compare the moment-resisting capacities of the sections shown, with appropriate inferences.

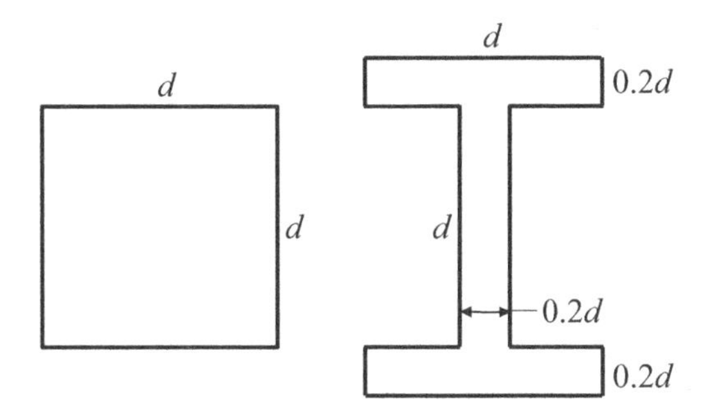

Solution steps:

(1) The section moduli need to be compared and are given by

$$Z_{\text{solid}} = \frac{\left(\frac{d^4}{12}\right)}{d/2} = \frac{d^3}{6}$$

$$Z_{\text{I-section}} = \frac{\left(\frac{d \times (1.4d)^3 - 0.8d \times d^3}{12}\right)}{0.7d} = 0.231d^3$$

(2) Thus, the moment-resisting capacity of the I-section is higher by a factor of $(6 \times 0.231) = 1.39$.

(3) Comparing the weights (i.e. cross-sectional areas), it is seen that the I-section weighs only 60% as much as the solid section. Thus, the I-section is a very efficient alternative to the solid section; this is because it has a more effective distribution of material—more

at the extreme fibre locations corresponding to maximum bending stress and less near the neutral axis.

Problem 7 Consider a cantilever of square cross-section subjected to uniformly distributed load q_o per unit length. With focus only on bending stresses, identify counterparts with a taper in (a) breadth alone; (b) depth alone; (c) breadth and depth simultaneously with both of them equal at any section, so as to achieve uniform maximum stress along the length. Compare their relative volumes.

Solution steps:

(1) With x measured from the tip as shown, $M(x) = -q_o x^2/2$; hence, for the uniform cantilever, the maximum extreme fibre stress occurs at the root ($x = L$) and rapidly decreases to zero at the tip. Thus, it is possible to taper the cross-sectional dimensions without affecting $q_{\text{allowable}}$.

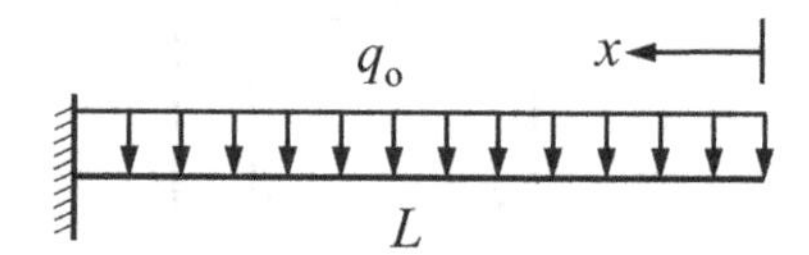

(2) At the root, taking the dimensions to be $b = d = d_o$,

$$|\sigma_{\max}| = \frac{\left(\frac{q_o L^2}{2}\right)\left(\frac{d_o}{2}\right)}{\left(\frac{d_o^4}{12}\right)} = \frac{3q_o L^2}{d_o^3}$$

(3) At any section x, with cross-sectional dimensions taken generically as $b(x)$ and $d(x)$,

$$|\sigma_{\max}| = \frac{\left(\frac{q_o x^2}{2}\right)\left(\frac{d(x)}{2}\right)}{\left(\frac{b(x)d(x)^3}{12}\right)} = \frac{3q_o x^2}{b(x)d(x)^2}$$

and this should be equated to the maximum stress at the root.

(4) For Case (a), putting $d(x) = d_o$, one gets $b(x) = d_o(x/L)^2$.

(5) For Case (b), putting $b(x) = d_o$, one gets $d(x) = d_o(x/L)$.

(6) For Case (c), putting $b(x) = d(x)$, one gets $d(x) = d_o(x/L)^{2/3}$.

(7) The variations are as shown.

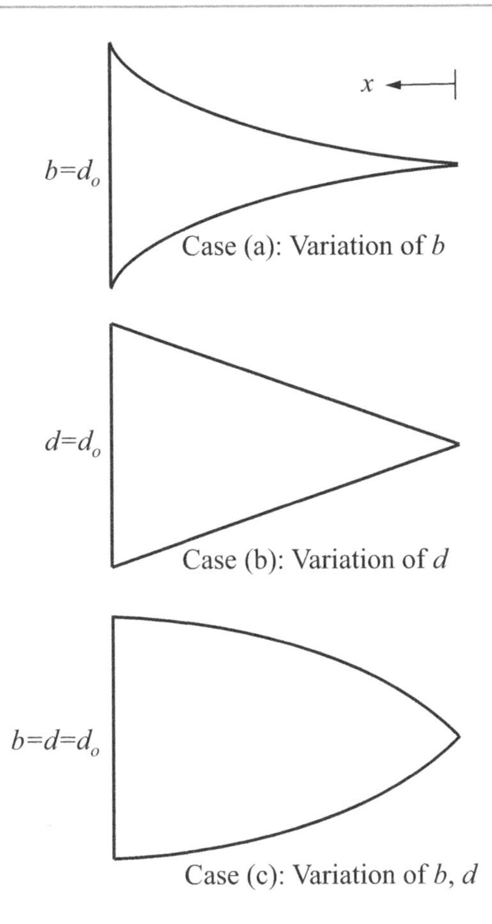

Case (a): Variation of *b*

Case (b): Variation of *d*

Case (c): Variation of *b*, *d*

(8) The volumes of each of these configurations may be calculated as integrals of $b(x)d(x)$ over the length; relative to the volume $d_o^2 L$ of the uniform beam, they turn out to be 1/3, 1/2, and 3/7, respectively, for Cases (a), (b), and (c).

Beams–Deflections 7

7.1 Engineering Beam Theory—Kinematics of Deformation

By virtue of the first and third assumptions of Sect. 6.1, the deformation of a small element of the beam is as in Fig. 7.1 which shows the kinematics on the longitudinal section (i.e. x–y plane) containing the centroidal axis.

Thus, the deformation of the entire beam may be described in terms of the radius of curvature $R(x)$, or the curvature $1/R(x)$, of the centroidal axis—with its sign convention corresponding to that adopted for the bending moment (see Fig. 5.4) and thus positive when the beam sags as in Fig. 7.1.

7.2 Deflection Analysis

At any point of the centroidal axis, the curvature is proportional to the bending moment. Using this fact and the expression for curvature in terms of the transverse deflection $v(x)$, the shape of the deformed axis (or the *elastic curve*) may be found out; further, the rotation θ of any element of the beam, approximated by the slope ($\tan\theta$) for small deformations, may be found as $v_{,x}$.

For such analysis, three methods are illustrated here—Double Integration Method, Moment-Area Method, and Superposition Method.

7.3 Final Formulae

$$\frac{E}{R} = \frac{M}{I} = \frac{\sigma_x}{y}; \quad \varepsilon_x = \frac{y}{R}$$

K. Bhaskar and T. K. Varadan, *Strength of Materials*,
https://doi.org/10.1007/978-3-031-06377-0_7

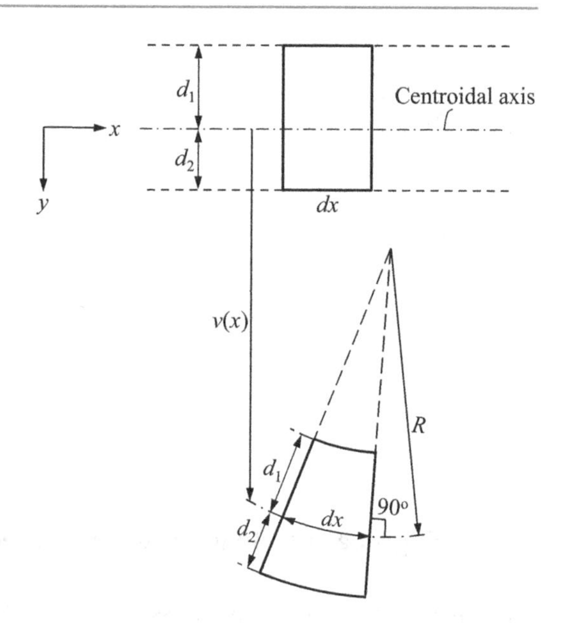

Fig. 7.1 Kinematics of beam bending

$$\frac{1}{R} = \frac{-v_{,xx}}{\left(1 + v_{,x}^2\right)^{3/2}} \approx -v_{,xx} \text{ for small rotations (see Sect. 1.3)}$$

$$EIv_{,xx} = -M$$

where $v(x)$ is positive downwards (i.e. in positive y-direction). It should be kept in mind that v and θ are continuous functions and cannot undergo sudden changes at any point; the curvature $v_{,xx}$, however, undergoes such sudden changes—at points where M or EI changes suddenly.

EI is called the flexural rigidity and is a measure of the resistance of the beam to bending deformation.

Using the relations between M, V and q (Sect. 5.3), one gets

$$\left(EIv_{,xx}\right)_{,x} = -V; \quad \left(EIv_{,xx}\right)_{,xx} = q$$

where the sign convention for $V(x)$ is as in Fig. 5.4 and $q(x)$, the load per unit length, is positive when acting downwards (in positive y direction).

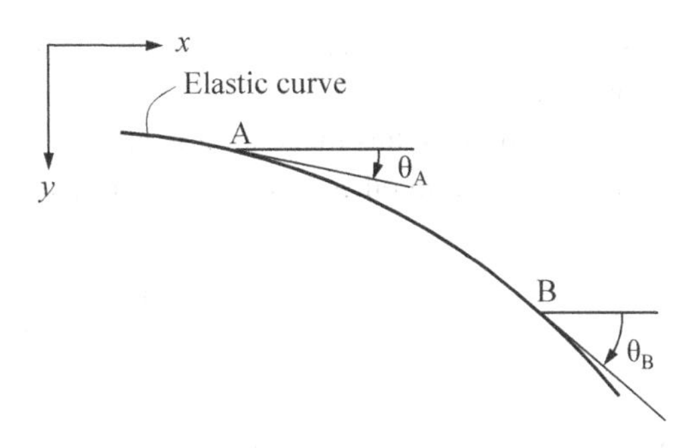

Fig. 7.2 Slopes at two points

Moment-Area Theorem I

$$\theta_B - \theta_A = -\left(\text{Area of } \frac{M}{EI} \text{ diagram between } A \text{ and } B\right)$$

where θ is the slope ($=v_{,x}$, positive when v increases with x, i.e. for a clockwise rotation from the undeformed position), and *point B is taken to the right of point A* (with a rightward x-coordinate as in Fig. 7.2). Further, the area of (M/EI) diagram between any two points is taken to be an algebraic quantity with the sign corresponding to M.

Moment-Area Theorem II

$$t_{B|A} = -\left(\text{Moment of } \frac{M}{EI} \text{ diagram between } A \text{ and } B \text{ }\underline{\text{about } B}\right)$$

where $t_{B|A}$ is the *tangential deviation* of point B with respect to point A, defined as the vertical distance of B from the tangent to the elastic curve (i.e. deformed centroidal axis) at A, and taken to be positive when B is below the tangent (Fig. 7.3). Once again, the area of (M/EI) diagram is taken to be an algebraic quantity as specified above while its

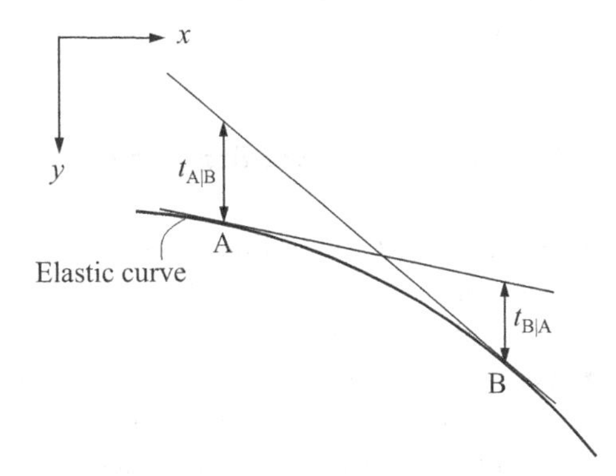

Fig. 7.3 Tangential deviations at two points

moment arm about B (i.e. distance of the centroid of the area from B) is always taken to be positive.

While applying this theorem, it is not necessary to take B to the right of A as was done with Theorem I; in other words, with reference to Fig. 7.3, it can also be used to calculate $t_{A|B}$ as

$$t_{A|B} = -\left(\text{Moment of } \frac{M}{EI} \text{ diagram between } A \text{ and } B \text{ } \underline{\text{about } A}\right)$$

where $t_{A|B}$ is the vertical distance of A from the tangent to the elastic curve at B, and positive when A is below the tangent. Note that $t_{A|B} \neq t_{B|A}$.

7.4 Specification of Boundary Conditions

The three commonly considered classical end conditions for the beam—clamped (also called fixed or restrained or built-in), simply supported (or hinged), and free—have already been introduced in Fig. 5.2. The difference between these conditions is in terms of either a complete suppression of the deflection or slope, or a complete absence of the corresponding restraining action, i.e. shear force or bending moment.

In mathematical terms, they are specified as

$$\begin{aligned}
&\text{Clamped end:} && v = v_{,x} = 0\\
&\text{Simply supported end:} && v = M = 0, \quad \text{i.e.} \quad v = v_{,xx} = 0\\
&\text{Free end:} && M = V = 0, \quad \text{i.e.} \quad v_{,xx} = \left(EIv_{,xx}\right)_{,x} = 0
\end{aligned}$$

For statically determinate beams, conditions corresponding to zero V or M are accounted for while drawing SFD and BMD while those in terms of v and $v_{,x}$ need to be enforced during deflection analysis.

For statically indeterminate beams, all the conditions need to be enforced together (see Sect. 7.6.4).

7.5 Range of Applicability

- As stated in Chap.6, Euler–Bernoulli hypothesis is acceptable only for long beams with length to depth ratio $\geq$20; for shorter beams, the actual deflections are significantly higher than the engineering beam theory estimates due to transverse shear deformation.
- For cantilevers and other beams with axially movable supports (e.g. roller-type simple supports), linear analysis is valid as long as the maximum deflections are small (one-tenth or less) compared to the length of the beam. Beyond this range, the expression

for curvature cannot be linearized by neglecting $v_{,x}^2$ compared to 1, and linear analysis leads to significant over-prediction of deflections.

- For beams with axially immovable supports, linear analysis is valid in a more restrictive range with the maximum deflections small (one-tenth or less) compared to the depth. Beyond this range, one has to account for the stiffening effect of axial stretching force induced by bending; due to this action, stresses across the depth will also be significantly affected.

7.6 Illustrative Problems

7.6.1 Double Integration Method

Based on integration of the moment–curvature relation, this method yields $v(x)$ valid over the entire length of the beam.

For problems where the expression for $M(x)$ is different for different segments of the beam, use of singularity functions is convenient as explained and illustrated in Problems 2, 3, and 4.

Problem 1 Find the equation of the elastic curve, and $v_{\max}$, $\theta_{\max}$ for Cases (i) and (iii). For Case (ii), find $v_{\max}$, $\theta_{\max}$ using the results of Case (i).

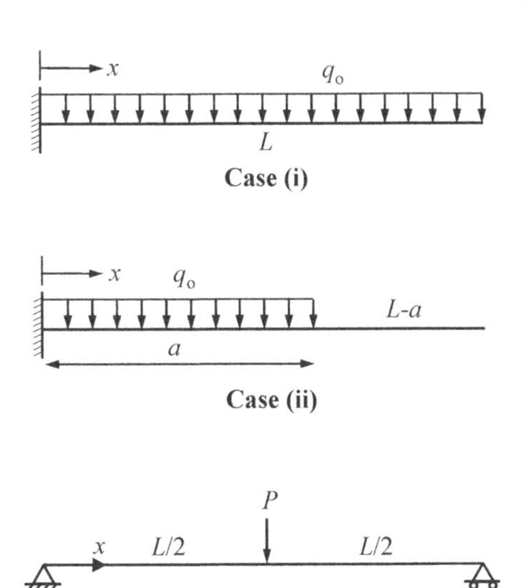

Solution steps for Case (i):

(1) $EIv_{,xx} = -M(x) = \frac{q_o(L-x)^2}{2}$

(2) $EIv_{,x} = -\frac{q_o(L-x)^3}{6} + C_1$

$EIv = \frac{q_o(L-x)^4}{24} + C_1 x + C_2$

(3) End conditions: $v = v_{,x} = 0$ at $x = 0 \rightarrow C_1 = \frac{q_o L^3}{6}, C_2 = -\frac{q_o L^4}{24}$

Thus, the elastic curve is given by

$$EIv(x) = \frac{q_o(L-x)^4}{24} + \frac{q_o L^3}{6}x - \frac{q_o L^4}{24}$$

(4) $\theta_{\max} = v_{,x}(L) = \frac{q_o L^3}{6EI}$; $v_{\max} = v(L) = \frac{q_o L^4}{8EI}$

Solution steps for Case (ii):

(1) Analysis of the loaded portion ($0 < x < a$) alone is exactly as done above and hence,

$$v_{,x}(a) = \frac{q_o a^3}{6EI};\ v(a) = \frac{q_o a^4}{8EI}$$

(2) The bending moment is zero for the unloaded right portion, and hence, it should remain straight; thus, the deformed beam is as shown below. (*Here, as well as in other such figures, the deflections are highly exaggerated for the sake of clarity.*)

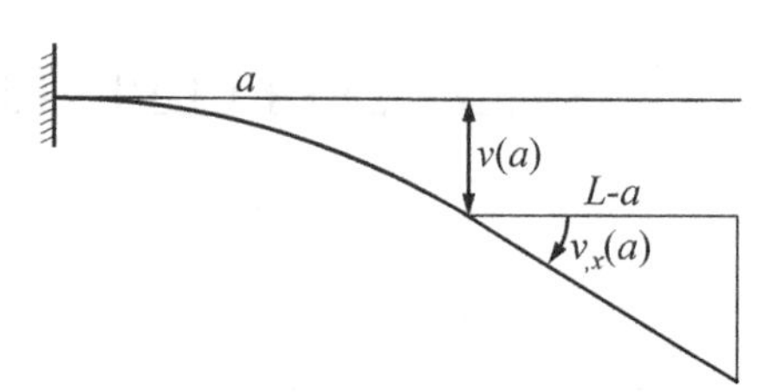

(3) $\theta_{\max} = v_{,x}(L) = v_{,x}(a) = \frac{q_o a^3}{6EI}$

$$v_{\max} = v(L) = v(a) + (L-a)v_{,x}(a)$$

$$= \frac{q_o a^4}{8EI} + \frac{(L-a)q_o a^3}{6EI} = \frac{q_o(4L-a)a^3}{24EI}$$

Solution steps for Case (iii):

(1) Due to symmetry, each support reaction = $P/2\uparrow$
(2) Elastic curve is symmetric about mid-span, and hence $v(x)$ may be found out for just one-half of the length. Further, at mid-span, slope is zero.
(3) For the left half: $M(x) = \frac{Px}{2}$
This, along with $v(0) = v_{,x}\left(\frac{L}{2}\right) = 0$, yields

$$EIv = \frac{PL^2x}{16} - \frac{Px^3}{12}; \quad \theta_{\max} = v_{,x}(0) = \frac{PL^2}{16EI}; \, v_{\max} = v\left(\frac{L}{2}\right) = \frac{PL^3}{48EI}$$

Problem 2 Find the elastic curve using singularity functions.

$2P$ $q_o = P/L$ P $L/4$ $L/4$ $L/2$ x

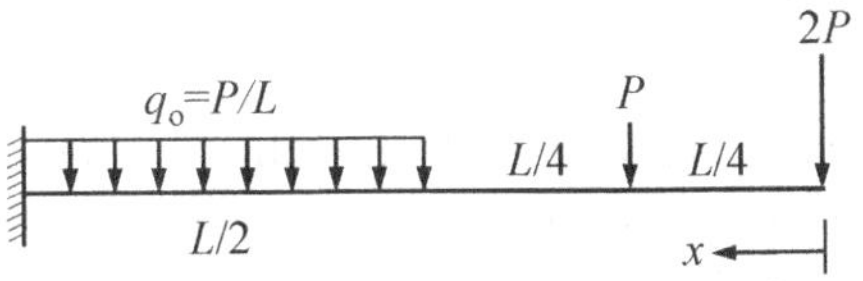

Solution steps:

(1) Taking x from the free end as shown, we have

$$M(x) = -2Px \quad \text{in the interval}\left(0, \frac{L}{4}\right)$$
$$= -2Px - P\left(x - \frac{L}{4}\right) \quad \text{in}\left(\frac{L}{4}, \frac{L}{2}\right)$$
$$= -2Px - P\left(x - \frac{L}{4}\right) - \frac{P\left(x - \frac{L}{2}\right)^2}{2L} \quad \text{in}\left(\frac{L}{2}, L\right)$$

Direct use of these in double integration method would lead to three different expressions for $v(x)$, one for each segment, along with six integration constants. To determine the constants, one has to impose the end conditions at $x = L$ along with continuity of v and $v_{,x}$ across the junctions of the segments at $x = L/4$ and $x = L/2$. Such tedious calculations may be obviated using singularity functions.

(2) A singularity function is defined as

$$\langle x - a\rangle = \begin{matrix} 0 & \text{for } x < a \\ (x - a) & \text{for } x \geq a \end{matrix}$$

and has the following properties.

$$\langle x-a\rangle^n = \begin{matrix} 0 & \text{for } x < a \\ 1 & \text{for } x \geq a, \quad n=0 \\ (x-a)^n & \text{for } x \geq a, \quad n>0 \end{matrix}$$

$$\frac{d\langle x-a\rangle^n}{dx} = n\langle x-a\rangle^{n-1} \text{ for } n \geq 1$$

$$\int \langle x-a\rangle^n dx = \frac{\langle x-a\rangle^{n+1}}{(n+1)} + c \quad \text{for } n \geq 0$$

Thus, for the present problem, the bending moment may be expressed in a unified form valid over the entire length as

$$M(x) = -2Px - P\left\langle x - \frac{L}{4}\right\rangle - \frac{P\left\langle x - \frac{L}{2}\right\rangle^2}{2L}$$

Note that the above expression is obtained by starting at $x = 0$ and considering moment contributions due to each load. (If one takes a rightward x-coordinate starting from the fixed end, then the first term will be the contribution due to the fixed end moment. Further, the uniform load will have to be continued beyond $x = L/2$ till the free end and a neutralizing opposite load introduced from $x = L/2$—see Problem 3).

(3) $EIv_{,x} = Px^2 + \frac{P\left\langle x-\frac{L}{4}\right\rangle^2}{2} + \frac{P\left\langle x-\frac{L}{2}\right\rangle^3}{6L} + C_1$

$EIv = \frac{Px^3}{3} + \frac{P\left\langle x-\frac{L}{4}\right\rangle^3}{6} + \frac{P\left\langle x-\frac{L}{2}\right\rangle^4}{24L} + C_1x + C_2$

Note that v and $v_{,x}$ are now automatically continuous across the junctions $x = L/4$ and $x = L/2$.

(4) The end conditions $v(L) = v_{,x}(L) = 0$ yield $C_1 = -\frac{125PL^2}{96}$, $C_2 = \frac{43PL^3}{48}$. The final expressions for $v(x)$ and $v_{,x}(x)$ are valid for all x; for example, they can be used to get

$$EIv_{\max} = EIv(0) = C_2 = \frac{43PL^3}{48};$$

$$EIv\left(\frac{L}{2}\right) = \left[\frac{P\left(\frac{L}{2}\right)^3}{3} + \frac{P\left(\frac{L}{2}-\frac{L}{4}\right)^3}{6} + C_1\left(\frac{L}{2}\right) + C_2\right] = \frac{37PL^3}{128};$$

$$EI\theta_{\max} = EIv_{,x}(0) = C_1 = -\frac{125PL^2}{96}$$ (minus indicating that v decreases with x, i.e. a clockwise rotation at the free end)

Problem 3 Find $v_{\max}$ and the end slopes.

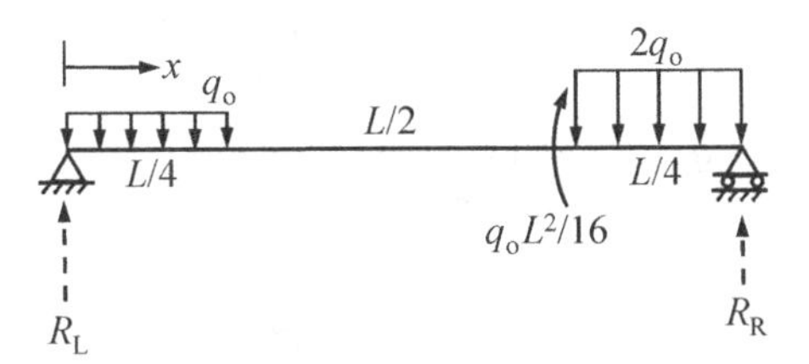

Solution steps:

(1) Moment about the right end $= 0 \rightarrow R_L = \frac{7q_oL}{32}$
(2) Taking x from the left end, one gets

$$
\begin{aligned}
M(x) &= \frac{7q_oLx}{32} - \frac{q_ox^2}{2} \text{ in}\left(0, \frac{L}{4}\right) \\
&= \frac{7q_oLx}{32} - \left(\frac{q_oL}{4}\right)\left(x - \frac{L}{8}\right) \text{ in}\left(\frac{L}{4}, \frac{3L}{4}\right) \\
&= \frac{7q_oLx}{32} - \left(\frac{q_oL}{4}\right)\left(x - \frac{L}{8}\right) + \frac{q_oL^2}{16} - \frac{2q_o\left(x - \frac{3L}{4}\right)^2}{2} \text{ in}\left(\frac{3L}{4}, L\right)
\end{aligned}
$$

Clearly, these cannot be expressed easily in a unified form since the second term in $M(x)$ is not the same for the first interval as for the next two intervals.

In order to get over this difficulty, it is necessary to continue the uniform load q_o till the end of the beam and to neutralize it over the second and third intervals by introducing an equal and opposite load starting at $x = L/4$ as shown below.

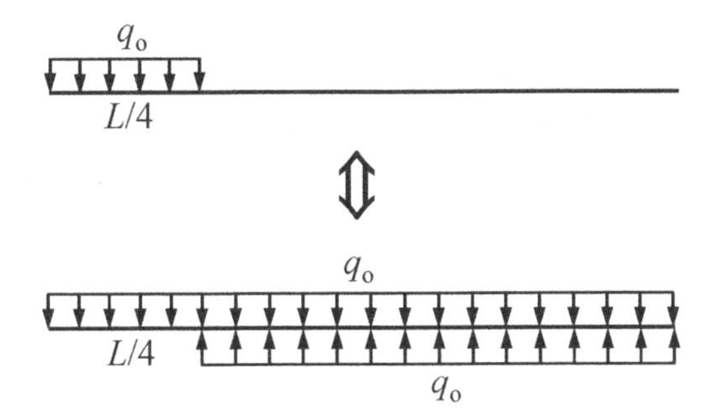

Thus, one gets

$$M(x) = \frac{7q_oLx}{32} - \frac{q_ox^2}{2} + \frac{q_o\left(x - \frac{L}{4}\right)^2}{2} \text{ in}\left(\frac{L}{4}, \frac{3L}{4}\right)$$
$$= \frac{7q_oLx}{32} - \frac{q_ox^2}{2} + \frac{q_o\left(x - \frac{L}{4}\right)^2}{2} + \frac{q_oL^2}{16} - \frac{2q_o\left(x - \frac{3L}{4}\right)^2}{2} \text{ in}\left(\frac{3L}{4}, L\right)$$

which are exactly equivalent to the earlier expressions but now enable a convenient restatement in terms of singularity functions as

$$M(x) = \frac{7q_oLx}{32} - \frac{q_ox^2}{2} + \frac{q_o\left\langle x - \frac{L}{4}\right\rangle^2}{2} + \frac{q_oL^2\left\langle x - \frac{3L}{4}\right\rangle^0}{16} - \frac{2q_o\left\langle x - \frac{3L}{4}\right\rangle^2}{2}$$

valid for all x.

(3) Proceeding further as in the earlier problems and applying the end conditions as $v(0) = v(L) = 0$, one gets

$$EIv = -\frac{7q_oLx^3}{192} + \frac{q_ox^4}{24} - \frac{q_o\left\langle x - \frac{L}{4}\right\rangle^4}{24}$$
$$-\frac{q_oL^2\left\langle x - \frac{3L}{4}\right\rangle^2}{32} + \frac{q_o\left\langle x - \frac{3L}{4}\right\rangle^4}{12} + \frac{59q_oL^3x}{6144}$$

$$v_{,x}(0) = \frac{59q_oL^3}{6144EI}\curvearrowright; \quad v_{,x}(L) = -\frac{85q_oL^3}{6144EI} = \frac{85q_oL^3}{6144EI}\curvearrowleft$$

(4) To find the location of $v_{\max}$, one has to check for the occurrence of zero slope in each segment. For the present problem, since the elastic curve is of fourth degree, it is more convenient to plot it and identify the location of $v_{\max}$ as $x \approx 0.53L$ and the corresponding value as $0.002693\frac{q_oL^4}{EI}$.

(It is important to note that the maximum deflection occurs close to mid-span even though the loading is unsymmetrical, and that one may get a very close estimate of it by considering the mid-span deflection itself—which is $0.002686\frac{q_oL^4}{EI}$ *in this case. This is in general true for unsymmetrically loaded simply supported beams except when a moment alone is applied.)*

Problem 4 Find the elastic curve.

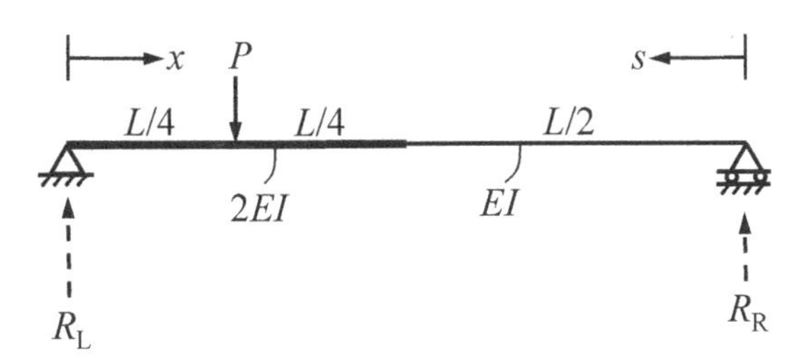

Solution steps:

(1) Calculation of support reactions as $R_L = \frac{3P}{4}$, $R_R = \frac{P}{4}$.
(2) Because of a change in EI at the center, a two-span solution is necessary.

$$M(x) = \frac{3Px}{4} - P\left\langle x - \frac{L}{4}\right\rangle \text{ for the left half;}$$

$$M(s) = \frac{Ps}{4} \text{ for the right half, with } x \text{ and } s \text{ taken as shown.}$$

(3) Double integration method leads to four undetermined constants.

The conditions

$v_L(x = 0) = v_R(s = 0) = 0$

$v_L\left(x = \frac{L}{2}\right) = v_R\left(s = \frac{L}{2}\right)$

$v_{L,x}\left(x = \frac{L}{2}\right) = -v_{R,s}\left(s = \frac{L}{2}\right)$(minus since x and s are in opposite directions)

finally yield

$$2EIv_L = -\frac{Px^3}{8} + \frac{P\left\langle x - \frac{L}{4}\right\rangle^3}{6} + \frac{25PL^2x}{384}$$

$$EIv_R = -\frac{Ps^3}{24} + \frac{23PL^2s}{768}$$

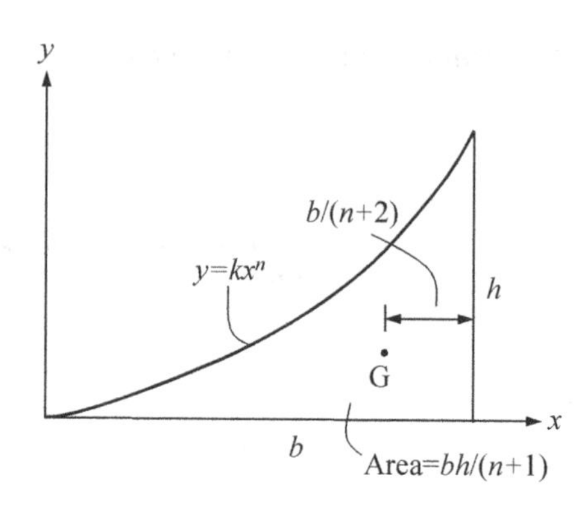

Fig. 7.4 Properties of area under kx^n curve

7.6.2 Moment-Area Method

This method is based on the use of BMD (often drawn by parts to facilitate calculations) to obtain slope and deflection at any desired specific point; it is considered most useful for stepped beams with piecewise constant *EI*. Its application is illustrated below for three categories of problems: cantilevers (with zero slope at the fixed end—Problems 5 and 6), simply supported beams under symmetric loading (with zero slope at mid-span—Problems 7 and 8) and those under unsymmetric loading (with slope unknown everywhere a priori—Problems 9 and 10).

For calculations of areas and moments of areas associated with BMD, it is convenient to note the general formulae pertaining to the area under the curve $y = kx^n$ as shown in Fig. 7.4.

Problem 5 For the cantilever of Problem 2, find the maximum slope and deflection.

Solution steps:

(1) The coordinate x is now taken from the left end. Point A is chosen at the fixed end where $\theta_A = 0$ and hence the tangent to the elastic curve is horizontal, and point B, to the right of A, at the free end. $\theta_{\max}$ and $v_{\max}$ occur at B, and $t_{B|A}$ is identified as relevant to v_B.

(2) BMD by parts as shown along with appropriate signs. Since *EI* is uniform, *M/EI* diagram not drawn separately.

(3) $\theta_B - \theta_A = \theta_B = -$(Area of *M/EI* diagram between A and B)

$$= -\left(\frac{1}{EI}\right)\left[-\frac{2PL.L}{2} - \frac{1}{2}\left(\frac{3PL}{4}\right)\left(\frac{3L}{4}\right) - \frac{1}{3}\left(\frac{PL}{8}\right)\left(\frac{L}{2}\right)\right] = \frac{125PL^2}{96EI} \curvearrowright$$

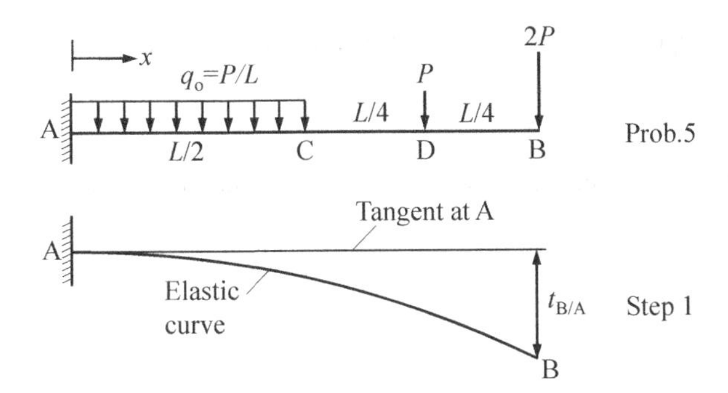

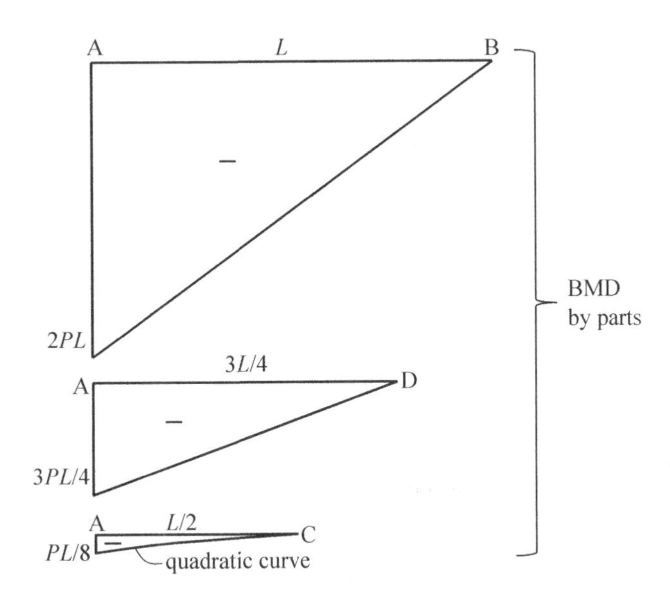

(4) $t_{B|A} = -$(Moment of *M/EI* diagram between *A* and *B* about B)

$$= -\left(\frac{1}{EI}\right)\left[-PL^2\left(\frac{2L}{3}\right) - \frac{9PL^2}{32}\left\{\frac{L}{4} + \frac{2}{3}\left(\frac{3L}{4}\right)\right\} - \frac{PL^2}{48}\left\{\frac{L}{2} + \frac{3}{4}\left(\frac{L}{2}\right)\right\}\right]$$

$$= \frac{43PL^3}{48EI}$$

$t_{B|A}$ is positive indicating B is below the tangent at A.
Hence, $v_B = \frac{43PL^3}{48EI} \downarrow$.

Problem 6 Repeat Problem 5 if the left half of the beam has flexural rigidity $2EI$.

Solution steps:

(1) BMD by parts as in Problem 5 and then *M/EI* diagram as shown.

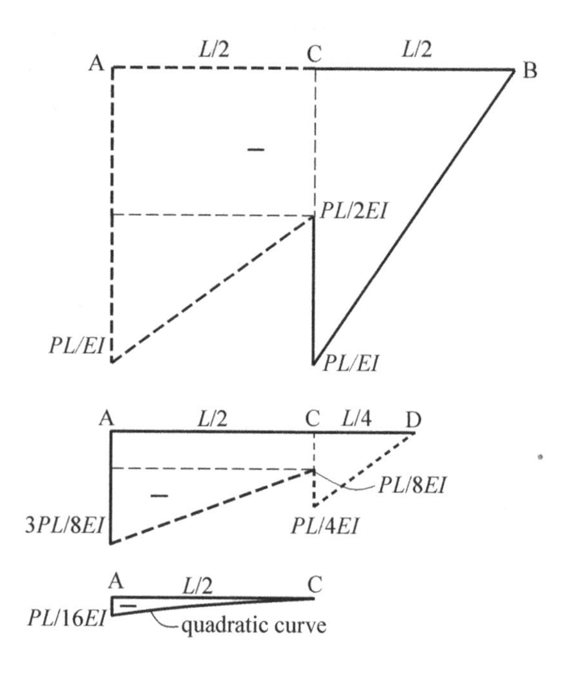

(2) Proceeding as before, and considering each trapezium as a rectangle plus a triangle,

$$\theta_B = -\begin{bmatrix} \{-\frac{1}{2}(\frac{PL}{EI})(\frac{L}{2}) - (\frac{PL}{2EI})(\frac{L}{2}) - \frac{1}{2}(\frac{PL}{2EI})(\frac{L}{2})\} + \\ \{-\frac{1}{2}(\frac{PL}{4EI})(\frac{L}{4}) - (\frac{PL}{8EI})(\frac{L}{2}) - \frac{1}{2}(\frac{PL}{4EI})(\frac{L}{2})\} + \\ \{-\frac{1}{3}(\frac{PL}{16EI})(\frac{L}{2})\} \end{bmatrix} = \frac{19PL^2}{24EI} \curvearrowright$$

$$t_{B|A} = -\begin{bmatrix} \left\{ \begin{matrix} -\left(\frac{PL^2}{4EI}\right)(\frac{2}{3} \cdot \frac{L}{2}) - \left(\frac{PL^2}{4EI}\right)(\frac{L}{2} + \frac{L}{4}) - \\ \left(\frac{PL^2}{8EI}\right)(\frac{L}{2} + \frac{2}{3} \cdot \frac{L}{2}) \end{matrix} \right\} + \\ \left\{ \begin{matrix} -\left(\frac{PL^2}{32EI}\right)(\frac{L}{4} + \frac{2}{3} \cdot \frac{L}{4}) - \left(\frac{PL^2}{16EI}\right)(\frac{L}{2} + \frac{L}{4}) - \\ \left(\frac{PL^2}{16EI}\right)(\frac{L}{2} + \frac{2}{3} \cdot \frac{L}{2}) \end{matrix} \right\} + \\ \left\{-\left(\frac{PL^2}{96EI}\right)(\frac{L}{2} + \frac{3}{4} \cdot \frac{L}{2})\right\} \end{bmatrix} = \frac{127PL^3}{256EI}$$

$$\Rightarrow v_B = \frac{127PL^3}{256EI} \downarrow$$

Problem 7 Sketch the elastic curve after calculating slopes and deflections at important points.

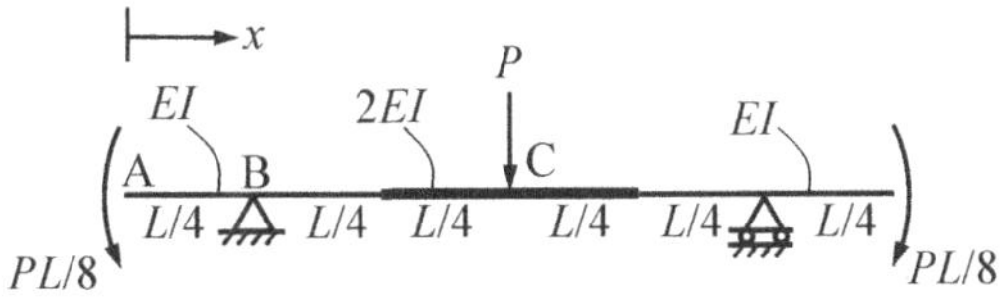

Solution steps:

(1) Each support reaction = $P/2 \uparrow$
(2) Due to symmetry about mid-span, the left half alone is considered with $\theta_C = 0$. BMD by parts and then *M/EI* diagram as shown.
(3) Elastic curve assumed as shown.
(4) θ_A is calculated using

$$\theta_C - \theta_A = -\theta_A = -\left(\text{Area of } \frac{M}{EI} \text{ diagram bet. A \& C}\right)$$

$$= -\left[\begin{array}{c}\left\{-\left(\frac{PL}{8EI}\right)\left(\frac{L}{2}\right) - \left(\frac{PL}{16EI}\right)\left(\frac{L}{4}\right)\right\} + \\ \left\{\frac{1}{2}\left(\frac{PL}{8EI}\right)\left(\frac{L}{4}\right) + \left(\frac{PL}{16EI}\right)\left(\frac{L}{4}\right) + \frac{1}{2}\left(\frac{PL}{16EI}\right)\left(\frac{L}{4}\right)\right\}\end{array}\right] = \frac{5PL^2}{128EI}$$

i.e. $\theta_A = -\frac{5PL^2}{128EI} = \frac{5PL^2}{128EI}$ ↶

(5) $\theta_B - \theta_A = -\left(\text{Area of } \frac{M}{EI} \text{ diagram bet. } A \ \& \ B\right)$

$$= -\left(-\frac{PL}{8EI}\right)\left(\frac{L}{4}\right) = \frac{PL^2}{32EI}$$

Thus, $\theta_B = -\frac{PL^2}{128EI} = \frac{PL^2}{128EI}$ ↶.

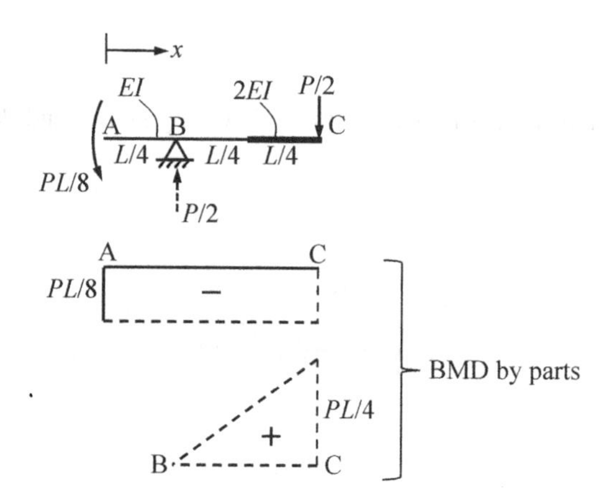

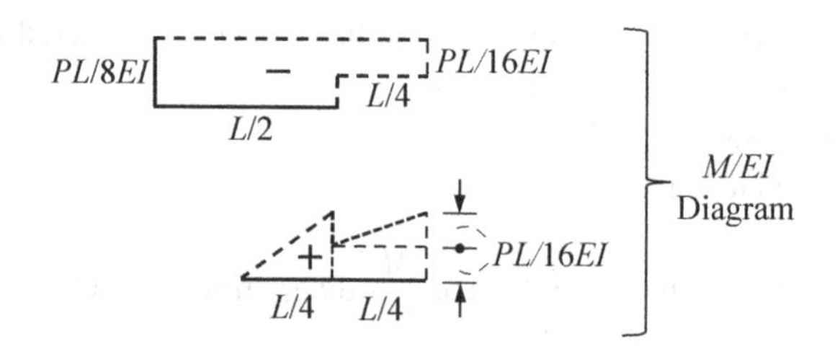

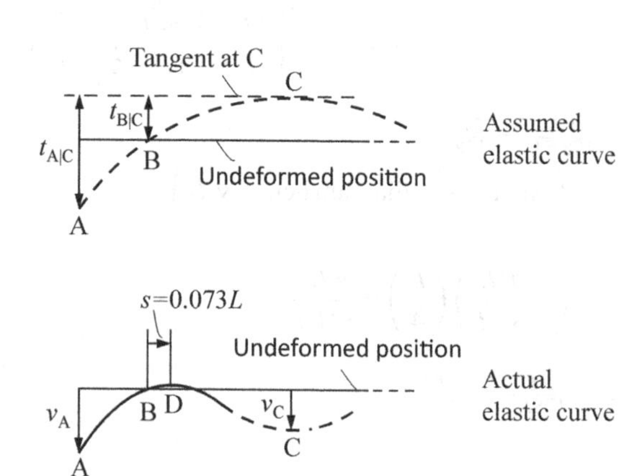

(6) With reference to the horizontal tangent at C,

$$t_{A|C} = -\left(\text{Moment of } \frac{M}{EI} \text{ diagram} \ldots \underline{\text{about } A}\right)$$

$$= -\left[\left\{\begin{array}{c}-\left(\frac{PL^2}{16EI}\right)\left(\frac{L}{4}\right)-\left(\frac{PL^2}{64EI}\right)\left(\frac{L}{2}+\frac{L}{8}\right)+\\ \left(\frac{PL^2}{64EI}\right)\left(\frac{L}{4}+\frac{2}{3}\cdot\frac{L}{4}\right)+\left(\frac{PL^2}{64EI}\right)\left(\frac{L}{2}+\frac{L}{8}\right)+\\ \left(\frac{PL^2}{128EI}\right)\left(\frac{L}{2}+\frac{2}{3}\cdot\frac{L}{4}\right)\end{array}\right\}\right] = \frac{PL^3}{256EI}$$

with the positive sign indicating that A is below the tangent at C.

Similarly,

$$t_{\mathrm{B|C}} = -\left(\text{Moment of } \frac{M}{EI} \text{ diagram} \ldots \underline{\text{about B}}\right) = -\frac{PL^3}{512EI}$$

with the negative sign indicating that B is above the tangent at C.
Hence, $v_C = \frac{PL^3}{512EI}$
and $v_A = \frac{PL^3}{512EI} + \frac{PL^3}{256EI} = \frac{3PL^3}{512EI}$,

both downwards from their original positions.

(7) Thus, the actual elastic curve is as shown.
The point D with zero slope just to the right of B may be located by using

$$\theta_D - \theta_B = \frac{PL^2}{128EI} = -(\text{Area} \ldots \text{bet. } B \;\&\; D)$$
$$= -\left[-\left(\frac{PL}{8EI}\right)s + \frac{1}{2}s\left(\frac{PL}{8EI}\right)\left(\frac{4s}{L}\right)\right]$$

to yield $s = 0.073L$ from B.
(The other root of the quadratic equation turns out to be $0.427L$ and is hence of no significance.)

Problem 8 Find the central deflection alone (see comments at the end of Problem 3).

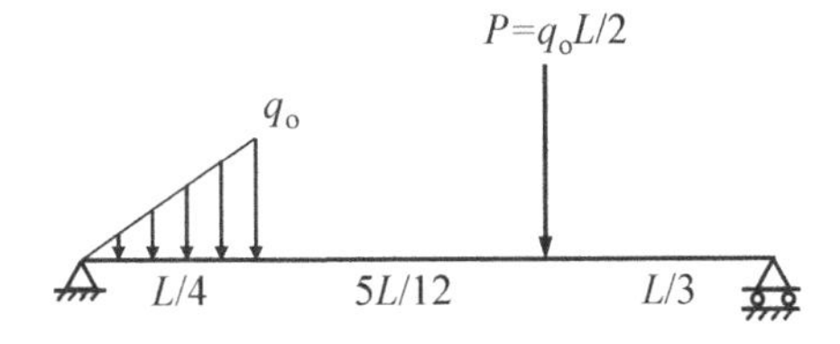

Solution steps:

(1) Since the central deflection would be the same if the applied loading is mirror-imaged about mid-span, one can consider the actual loading and its mirror image together as shown and obtain the required answer by halving the resulting central deflection. Thus, the problem is reduced to one of symmetric loading with support reactions as shown.

(2) Further, to facilitate moment-area calculations, the triangular load is extended till mid-span and neutralized by a trapezoidal load as shown.

(3) BMD by parts as shown, with the trapezoidal load looked upon as uniform load plus a triangular load.

(4) With reference to the horizontal tangent at C,

$$t_{A|C} = -\frac{1}{EI}(\text{Moment of BMD} \ldots \underline{\text{about A}})$$

$$= -\frac{1}{EI}\left[\begin{array}{c}\left(\frac{1}{2}\cdot\frac{5q_oL^2}{16}\cdot\frac{L}{2}\right)\left(\frac{2}{3}\cdot\frac{L}{2}\right) - \left(\frac{1}{4}\cdot\frac{q_oL^2}{12}\cdot\frac{L}{2}\right)\left(\frac{4}{5}\cdot\frac{L}{2}\right) \\ +\left(\frac{1}{3}\cdot\frac{q_oL^2}{32}\cdot\frac{L}{4}\right)\left(\frac{L}{4}+\frac{3}{4}\cdot\frac{L}{4}\right) + \left(\frac{1}{4}\cdot\frac{q_oL^2}{96}\cdot\frac{L}{4}\right)\left(\frac{L}{4}+\frac{4}{5}\cdot\frac{L}{4}\right) \\ -\left(\frac{1}{2}\cdot\frac{q_oL^2}{12}\cdot\frac{L}{6}\right)\left(\frac{L}{3}+\frac{2}{3}\cdot\frac{L}{6}\right)\end{array}\right]$$

$$= -0.02022\frac{q_oL^4}{EI}$$

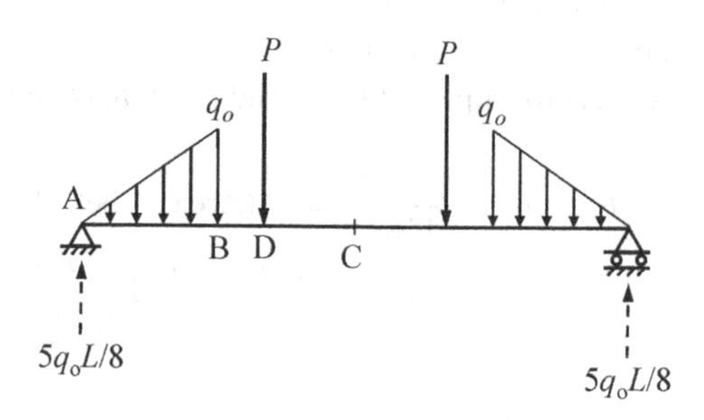

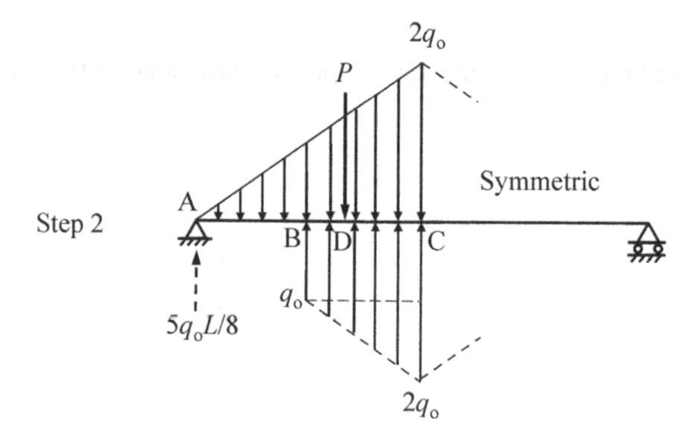

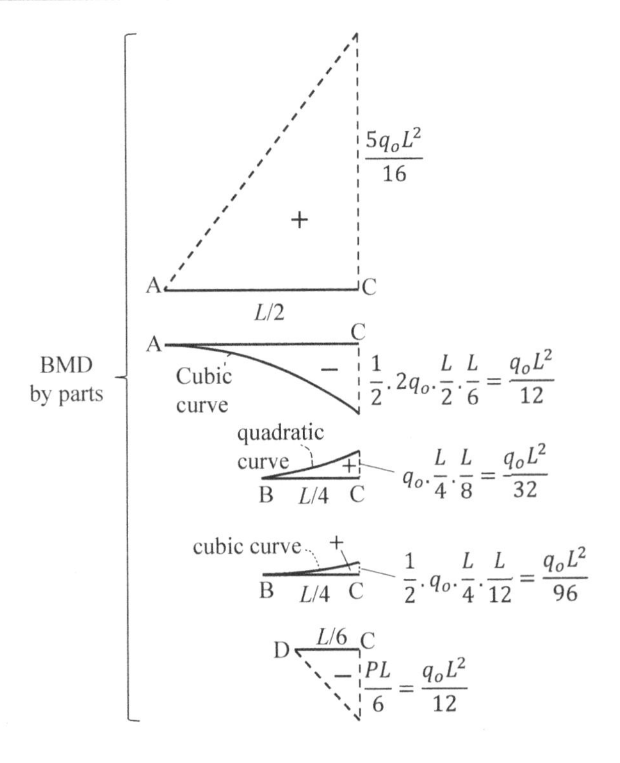

with the minus sign implying that A is above the tangent at C.

(5) Hence, v_C due to actual load $= 0.01011\frac{q_oL^4}{EI}\downarrow$

Problem 9 Find the slopes at the two ends and the central deflection.

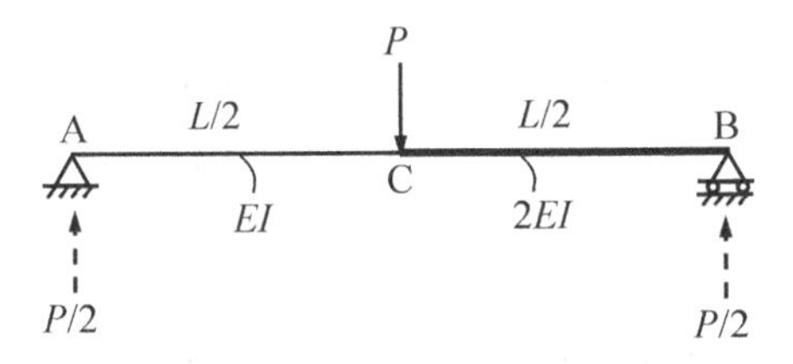

Solution steps:

(1) BMD and (M/EI) diagrams as shown.

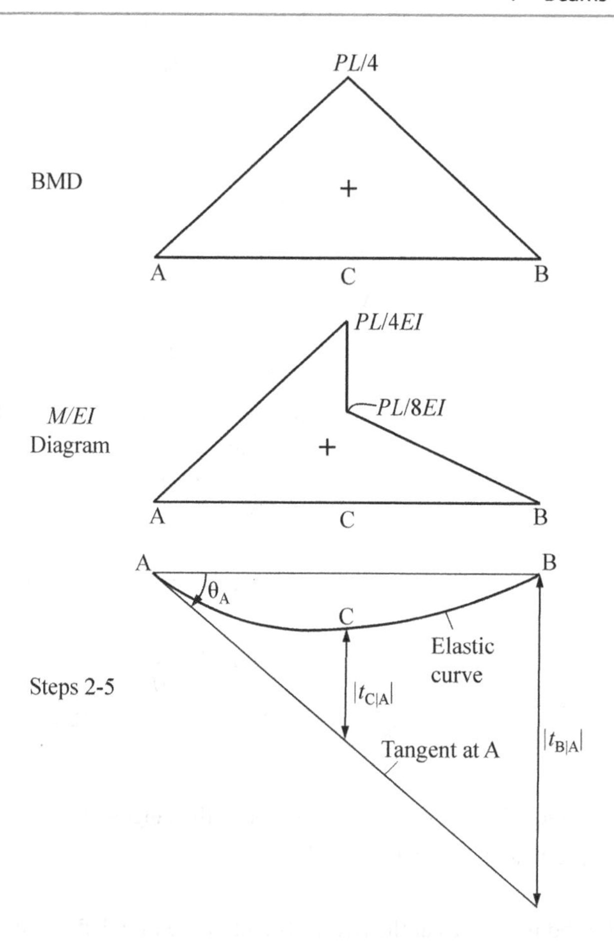

(2) With reference to the tangent drawn at A,

$$t_{B|A} = -\left(\text{Moment of } \frac{M}{EI} \text{ diagram} \ldots \underline{\text{about } B}\right)$$

$$= -\left[\left(\frac{1}{2} \cdot \frac{PL}{4EI} \cdot \frac{L}{2}\right)\left(\frac{L}{2} + \frac{1}{3} \cdot \frac{L}{2}\right) + \left(\frac{1}{2} \cdot \frac{PL}{8EI} \cdot \frac{L}{2}\right)\left(\frac{2}{3} \cdot \frac{L}{2}\right)\right] = -\frac{5PL^3}{96EI}$$

where the minus sign implies that B is above the tangent as shown.

(3) Since $|t_{B|A}|$ is also equal to $L\theta_A$, one gets $\theta_A = \frac{5PL^2}{96EI} \curvearrowright$

(4) With θ_A known,

$$\theta_B - \theta_A = -\left(\text{Area of } \frac{M}{EI} \text{ diagram}\right)$$
$$= -\left(\frac{1}{2}.\frac{PL}{4EI}.\frac{L}{2} + \frac{1}{2}.\frac{PL}{8EI}.\frac{L}{2}\right) = -\frac{3PL^2}{32EI}$$

yielding $\theta_B = -\frac{PL^2}{24EI} = \frac{PL^2}{24EI}$ ↶.

(5) The deflection at C is calculated as the difference between its undeformed vertical position with reference to the tangent at A, given by $(\frac{L}{2})\theta_A$, and its deformed position given by the magnitude of $t_{C|A}$.

$$t_{C|A} = -\left(\text{Moment of } \frac{M}{EI} \text{ diagram} \ldots \underline{\text{about } C}\right)$$
$$= -\left(\frac{1}{2}.\frac{PL}{4EI}.\frac{L}{2}\right)\left(\frac{1}{3}.\frac{L}{2}\right) = -\frac{PL^3}{96EI}$$

which is negative since C is above the tangent at A.
Hence, $v_C = \left(\frac{L}{2}\cdot\frac{5PL^2}{96EI}\right) - \frac{PL^3}{96EI} = \frac{PL^3}{64EI}$ ↓.
(This result may also be obtained by similar steps with reference to the tangent at B.)

Problem 10 Find v_A.

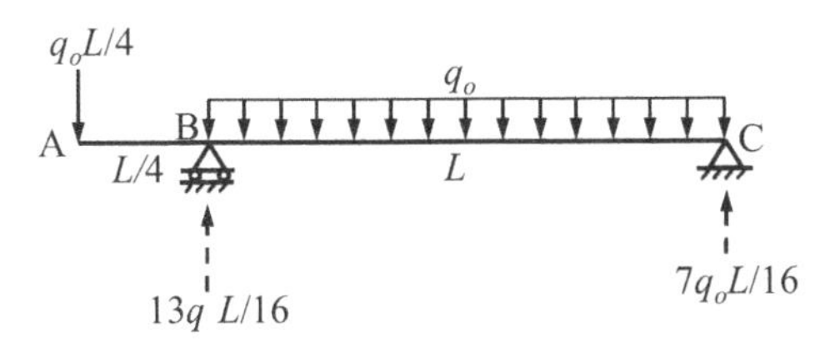

Solution steps:

(1) BMD by parts as shown.
(2) Assuming the elastic curve as shown,

$$t_{C|B} = -\frac{1}{EI}\left(\text{Moment of BMD} \ldots \underline{\text{about } C}\right)$$
$$= -\frac{1}{EI}\left(-\frac{q_oL^2}{16}.L.\frac{L}{2} - \frac{1}{2}.\frac{q_oL^2}{4}.L.\frac{L}{3} + \frac{1}{2}.\frac{13q_oL^2}{16}.L.\frac{L}{3} - \frac{1}{3}.\frac{q_oL^2}{2}.L.\frac{L}{4}\right)$$
$$= -\frac{q_oL^4}{48EI}$$

with the minus sign indicating that C is indeed above the tangent as assumed. Hence, $\theta_B = \frac{|t_{C|B}|}{L} = \frac{q_o L^3}{48EI}\curvearrowright$.

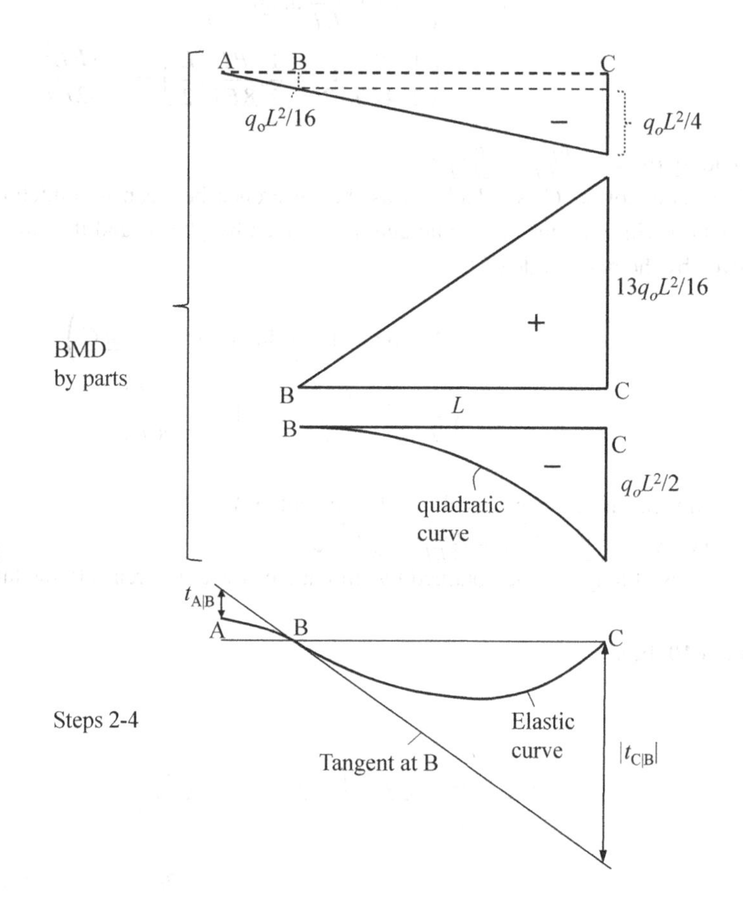

(3) Thus the undeformed position of A is $\left(\frac{L}{4}\right)\theta_B$ below the tangent at B, while its deformed position is given by

$$t_{A|B} = -\frac{1}{EI}\left(\text{Moment of BMD} \ldots \underline{\text{about A}}\right)$$

$$= -\frac{1}{EI}\left(-\frac{1}{2}.\frac{q_o L^2}{16}.\frac{L}{4}\right).\left(\frac{2}{3}.\frac{L}{4}\right) = \frac{q_o L^4}{768EI}$$

with the positive value indicating that A is below the tangent.

(4) Hence, $v_A = \frac{q_o L^4}{768EI} - \frac{L}{4}.\frac{q_o L^3}{48EI} = -\frac{q_o L^4}{256EI}$

with the minus sign indicating an upward deflection as assumed.

7.6.3 Superposition Method

For beams undergoing small elastic deformations due to several transverse loads applied together, the deflection or slope at any point may be obtained by considering the effect of each load separately and subsequent algebraic summation; this is possible since the differential equation governing small deformations is linear and so are the boundary conditions. Thus, a tabulated database of deflection and slope results for some basic cases (as in Table 7.1) may readily be used and the results for any given loading may be

Table 7.1 Deflection and slope results for some basic cases (v positive if downwards; θ positive if clockwise; x rightward from A)

1	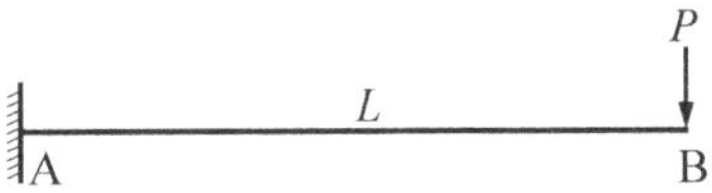 Cantilever fixed at A, length L, point load P at free end B	$\theta_B = \frac{PL^2}{2EI}$ $v_B = \frac{PL^3}{3EI}$
2	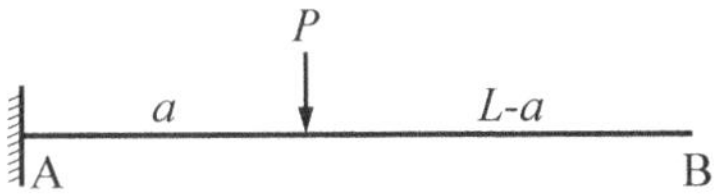 Cantilever fixed at A, point load P at distance a from A, $L-a$ to B	$\theta_B = \frac{Pa^2}{2EI}$ $v_B = \frac{Pa^2}{6EI}(3L-a)$
3	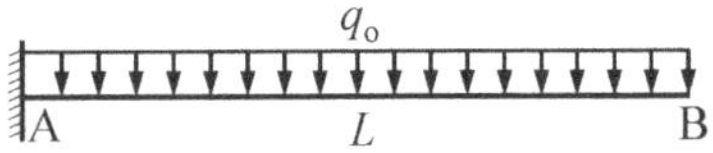 Cantilever fixed at A, uniform load q_o over length L	$\theta_B = \frac{q_oL^3}{6EI}$ $v_B = \frac{q_oL^4}{8EI}$
4	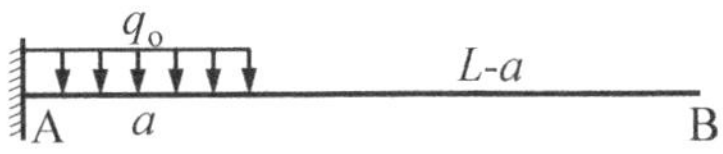 Cantilever fixed at A, uniform load q_o over length a from A, $L-a$ unloaded	$\theta_B = \frac{q_oa^3}{6EI}$ $v_B = \frac{q_oa^3}{24EI}(4L-a)$
5	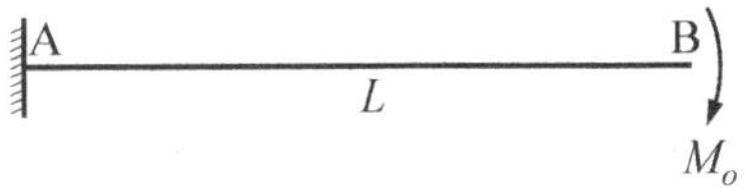 Cantilever fixed at A, length L, moment M_o at B	$\theta_B = \frac{M_oL}{EI}$ $v_B = \frac{M_oL^2}{2EI}$
6	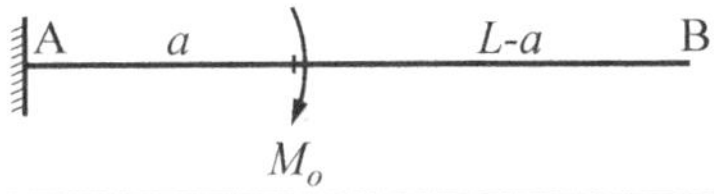 Cantilever fixed at A, moment M_o at distance a from A, $L-a$ to B	$\theta_B = \frac{M_oa}{EI}$ $v_B = \frac{M_oa}{2EI}(2L-a)$
7	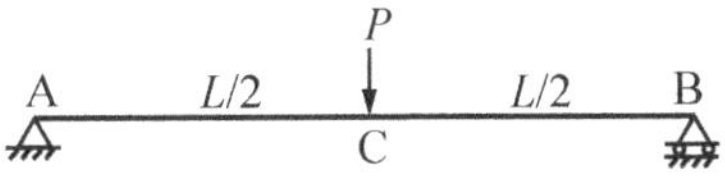 Simply supported at A and B, point load P at midspan C ($L/2$, $L/2$)	$\theta_A = -\theta_B = \frac{PL^2}{16EI}$ $v_C = \frac{PL^3}{48EI}$

(continued)

Table 7.1 (continued)

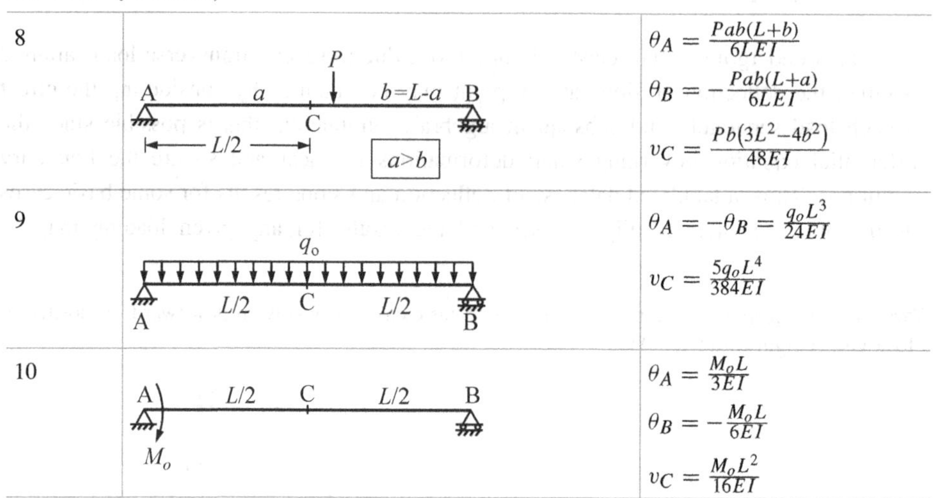

8	A, a, P, b=L-a, B, C, $L/2$, a>b	$\theta_A = \frac{Pab(L+b)}{6LEI}$ $\theta_B = -\frac{Pab(L+a)}{6LEI}$ $v_C = \frac{Pb(3L^2-4b^2)}{48EI}$
9	q_o, $L/2$, C, $L/2$, A, B	$\theta_A = -\theta_B = \frac{q_oL^3}{24EI}$ $v_C = \frac{5q_oL^4}{384EI}$
10	A, $L/2$, C, $L/2$, B, M_o	$\theta_A = \frac{M_oL}{3EI}$ $\theta_B = -\frac{M_oL}{6EI}$ $v_C = \frac{M_oL^2}{16EI}$

obtained by appropriate superposition. It should be noted that several alternative load–case combinations are possible for any given problem and the actual choice depends on one's convenience.

Problem 11 Find the maximum deflection and slope.

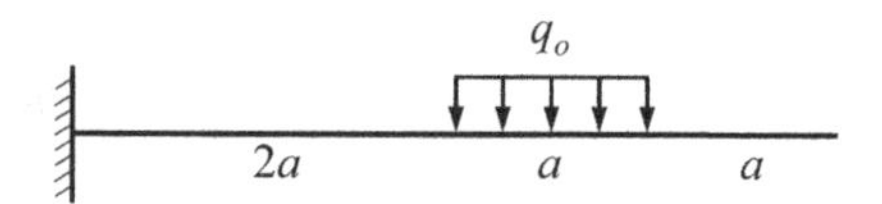

Solution steps:

(1) Look at this problem as subtraction of Case (b) from Case (a) as shown.

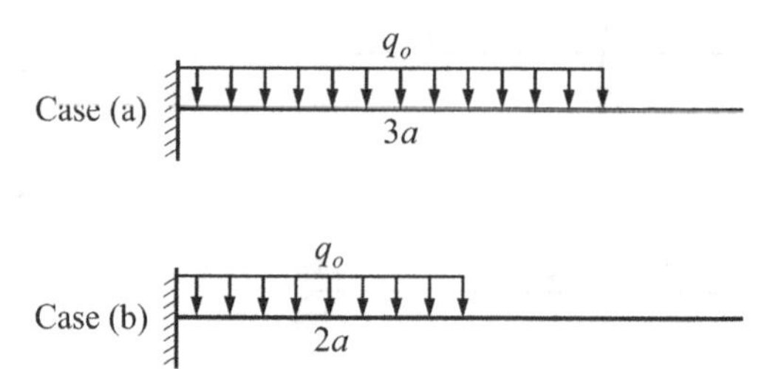

(2) Then, one gets from Table 7.1, S.No. 4,

$$v_{\max} = \frac{q_o(3a)^3}{24EI}\{4(4a) - 3a\} - \frac{q_o(2a)^3}{24EI}\{4(4a) - 2a\} = \frac{239q_oa^4}{24EI} \downarrow$$

Similarly, noting that $\theta_{\max}$ due to the actual load occurs at the free end, one gets it as

$$\theta_{\max} = \frac{q_o(3a)^3}{6EI} - \frac{q_o(2a)^3}{6EI} = \frac{19q_oa^3}{6EI} \curvearrowright$$

(3) As an alternative to the above solution, one may look at the applied loading as a series of concentrated loads acting over infinitesimally small lengths dx, and use the results of Table 7.1, S.No. 2 for any such load located at x from the left end by taking $a = x$; the net effect is obtained by integration from $x = 2a$ to $3a$.

Thus, $v_{\max} = \int_{2a}^{3a} \frac{q_o x^2}{6EI}[3(4a) - x]dx = \frac{239q_oa^4}{24EI} \downarrow$

$$\theta_{\max} = \int_{2a}^{3a} \frac{q_o x^2}{2EI}dx = \frac{19q_oa^3}{6EI} \curvearrowright$$

Problem 12 Find θ_A, θ_C, v_B, and v_D.

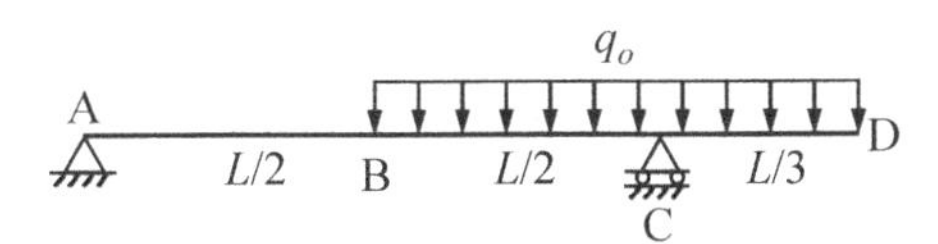

Solution steps:

(1) Consider portion AC first—its deformation is due to the action of two load cases as shown, with Case (a) corresponding to load q_o applied over BC alone and Case (b) to that applied over CD alone.

(2) For Case (a), one can consider the distributed load as a series of concentrated loads and use the results of Table 7.1, S.No. 8 for any such load located at x from the left end by taking $a = x$, $b = L\text{-}x$.

While this method is required for the two slopes, v_B may be found out more easily as half of that due to q over the entire length (Table 7.1, S.No. 9).

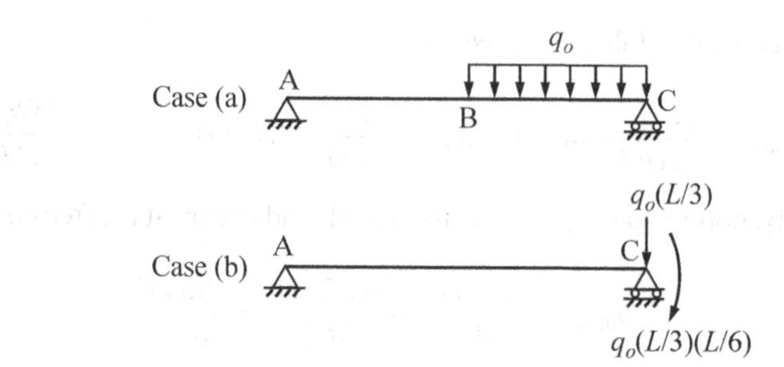

(3) Thus,

$$\theta_A = \int_{\frac{L}{2}}^{L} \frac{q_o x(L-x)(L+L-x)}{6LEI} dx = \frac{7q_oL^3}{384EI}$$

$$\theta_C = -\int_{\frac{L}{2}}^{L} \frac{q_o x(L-x)(L+x)}{6LEI} dx = -\frac{3q_oL^3}{128EI}; \quad v_B = \frac{1}{2}\left(\frac{5q_oL^4}{384EI}\right)$$

(4) For Case (b), the downward force at C does not cause any deflection because it is resisted directly by the support.

Due to the moment, from (Table 7.1, S.No.10) with appropriate sign changes, $\theta_A = -\left(\frac{q_oL^2}{18}\right)\left(\frac{L}{6EI}\right)$; $\theta_C = \left(\frac{q_oL^2}{18}\right)\left(\frac{L}{3EI}\right)$; $v_B = -\left(\frac{q_oL^2}{18}\right)\left(\frac{L^2}{16EI}\right)$

(5) Thus, the net values are

$$\theta_A = \frac{31q_oL^3}{3456EI} \curvearrowright; \quad \theta_C = -\frac{17q_oL^3}{3456EI} = \frac{17q_oL^3}{3456EI} \curvearrowleft; \quad v_B = \frac{7q_oL^4}{2304EI} \downarrow$$

(6) To find v_D, it is necessary to superpose two displacements—one due to rotation of CD as a rigid body about C by θ_C, and the other due to deformation of CD as a cantilever fixed at C and subjected to uniform load (Table 7.1, S.No. 3). The corresponding values may simply be added together algebraically since the deflections are small.

Thus, $v_D = \theta_C\left(\frac{L}{3}\right) + \frac{q_o}{8EI}\left(\frac{L}{3}\right)^4 = -\frac{q_oL^4}{10368EI}$

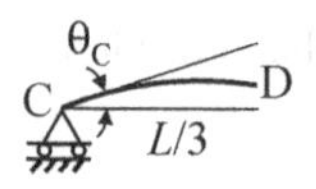

Fig. 7.5 Statically indeterminate beams

(a) Propped cantilever

(b) Clamped beam

(c) Continuous beam

where the minus sign indicates that D moves upwards as shown.

7.6.4 Statically Indeterminate Beams

These are beams with extra, redundant restraints beyond those essential to arrest rigid body motions; they include propped cantilevers, beams clamped at both ends, and continuous beams with three or more supports wherein the outer ones may be simple supports (with possible overhangs) or clamped ends (Fig. 7.5).

These beams may be analysed by either superposition method or double integration method as shown below.

Problem 13 Find the prop reaction and θ_B using superposition method.

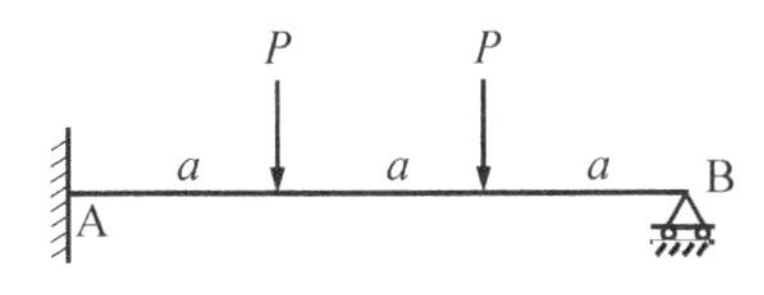

Solution steps:

(1) The problem may be considered as that of a simple statically determinate cantilever with the redundant support reaction as an unknown additional loading besides the applied external loads; thus, it may be reduced to superposition of the load cases shown. (Alternatively, one may consider a simply supported beam with a left-end moment superposed along with the two point loads.)

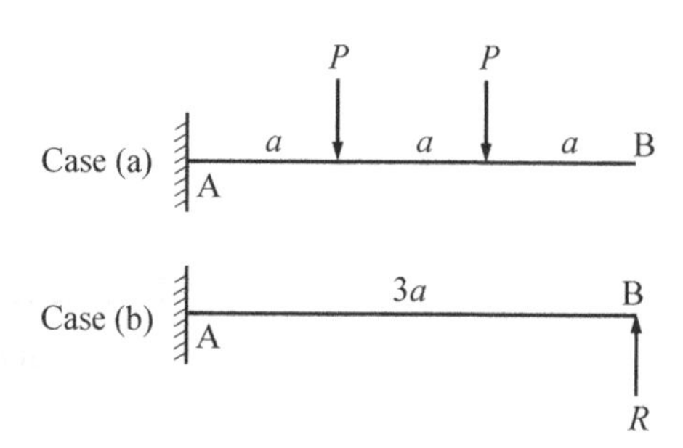

(2) For Case (a), using Table 7.1, S.No. 2 for each point load, one gets

$$v_B = \frac{Pa^2}{6EI}(9a - a) + \frac{P(2a)^2}{6EI}(9a - 2a) = \frac{6Pa^3}{EI}$$

$$\theta_B = \frac{Pa^2}{2EI} + \frac{P(2a)^2}{2EI} = \frac{5Pa^2}{2EI}$$

(3) For Case (b), from Table 7.1, S.No. 1,

$$v_B = -\frac{R(3a)^3}{3EI}; \quad \theta_B = -\frac{R(3a)^2}{2EI}$$

(4) Net $v_B = 0 \Rightarrow R = \frac{2P}{3}$

(5) Hence, for the propped cantilever,

$$\theta_B = \frac{5Pa^2}{2EI} - \left(\frac{2P}{3}\right)\frac{(3a)^2}{2EI} = -\frac{Pa^2}{2EI} = \frac{Pa^2}{2EI}\curvearrowleft$$

Problem 14 Obtain the elastic curve for the above problem using double integration method.

Solution steps:

(1) With a leftward x-coordinate from the prop,

$$M(x) = Rx - P\langle x - a\rangle - P\langle x - 2a\rangle$$

(2) $EIv_{,x} = \frac{-Rx^2}{2} + \frac{P\langle x-a\rangle^2}{2} + \frac{P\langle x-2a\rangle^2}{2} + C_1$

$$EIv = \frac{-Rx^3}{6} + \frac{P\langle x - a\rangle^3}{6} + \frac{P\langle x - 2a\rangle^3}{6} + C_1x + C_2$$

(3) The three unknowns R, C_1, and C_2 may be determined by imposing the conditions $v(0) = v(3a) = v_{,x}(3a) = 0$. Thus, one gets

$$EIv = \frac{-Px^3}{9} + \frac{P\langle x-a\rangle^3}{6} + \frac{P\langle x-2a\rangle^3}{6} + \frac{Pa^2x}{2}$$

Problem 15 Find the support reactions.

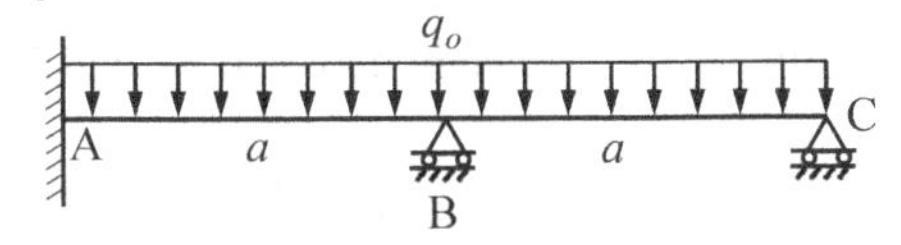

Solution steps:

(1) Consider this problem as superposition of the load cases shown for a simply supported beam.

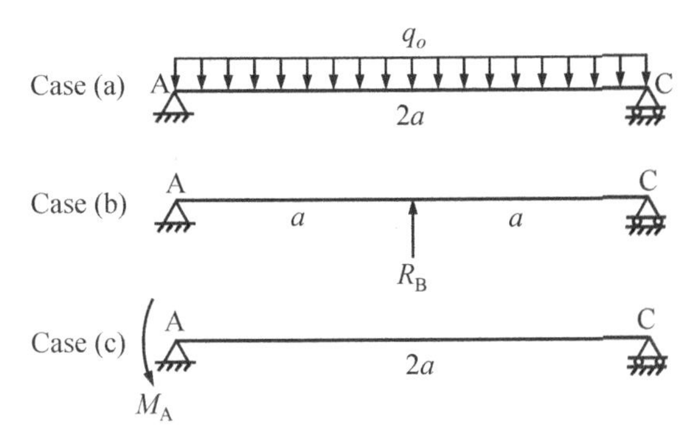

(2) Using the results of Table 7.1, S.Nos. 9,7 and 10, we get

$$\theta_A = \frac{q_o(2a)^3}{24EI} - \frac{R_B(2a)^2}{16EI} - \frac{M_A(2a)}{3EI} = 0$$

$$v_B = \frac{5q_o(2a)^4}{384EI} - \frac{R_B(2a)^3}{48EI} - \frac{M_A(2a)^2}{16EI} = 0$$

(3) These equations yield $R_B = \frac{8q_oa}{7} \uparrow$, $M_A = \frac{q_oa^2}{14} \curvearrowleft$

(4) The other support reactions are obtained by superposition of those from Cases (a)–(c) as

$$R_A = q_oa - \frac{R_B}{2} + \frac{M_A}{2a} = \frac{13q_oa}{28} \uparrow$$

$$R_C = q_oa - \frac{R_B}{2} - \frac{M_A}{2a} = \frac{11q_oa}{28} \uparrow$$

Problem 16 For the above problem, find the elastic curve using double integration method.

Solution steps:

(1) With a leftward x-coordinate from the right end, and R_B and R_C taken to be upwards,

$$-EIv_{,xx} = M(x) = R_C x - \frac{q_o x^2}{2} + R_B \langle x - a \rangle$$

(2) Hence,

$$EIv_{,x} = -\frac{R_C x^2}{2} + \frac{q_o x^3}{6} - \frac{R_B \langle x - a \rangle^2}{2} + C_1$$

$$EIv = -\frac{R_C x^3}{6} + \frac{q_o x^4}{24} - \frac{R_B \langle x - a \rangle^3}{6} + C_1 x + C_2$$

(3) Using $v(0) = v(a) = v(2a) = v_{,x}(2a) = 0$, one gets the unknowns as

$$R_B = \frac{8q_o a}{7}, \quad R_C = \frac{11q_o a}{28}, \quad C_1 = \frac{q_o a^3}{42}, \quad C_2 = 0$$

(4) Thus, the elastic curve is given by

$$EIv = -\frac{11q_o a x^3}{168} + \frac{q_o x^4}{24} - \frac{4q_o a \langle x - a \rangle^3}{21} + \frac{q_o a^3 x}{42}$$

Combined Loading, Stress Transformation, Failure Criteria

8

8.1 Combined Loading

Using the principle of superposition, a structure subjected to combined loading is analysed by considering one load at a time and then the net effect; this procedure results in identification of one or more critical points at which the state of stress is severe and may lead to failure.

Identification of critical points and determination of the corresponding states of stress is first illustrated by way of the following problems wherein the notation adopted in earlier chapters is adhered to as far as possible; it should be noted that the state of stress turns out to be two-dimensional in all the problems.

Problem 1: A walking stick is subjected to an eccentric load as shown. Find the high-stress location and the state of stress.

Solution steps:

(1) Free body diagram (FBD) of vertical portion as shown. All cross-sections along the length are loaded identically.

(Note that an additional bending moment, equal to Pv, occurs at any cross-section if one considers the deformed configuration as shown. This second-order effect may be neglected as long as P is very small compared to the buckling load of the stick, and such a situation is assumed here. For further details, see Chap. 9, Problem 2.)

K. Bhaskar and T. K. Varadan, *Strength of Materials*,
https://doi.org/10.1007/978-3-031-06377-0_8

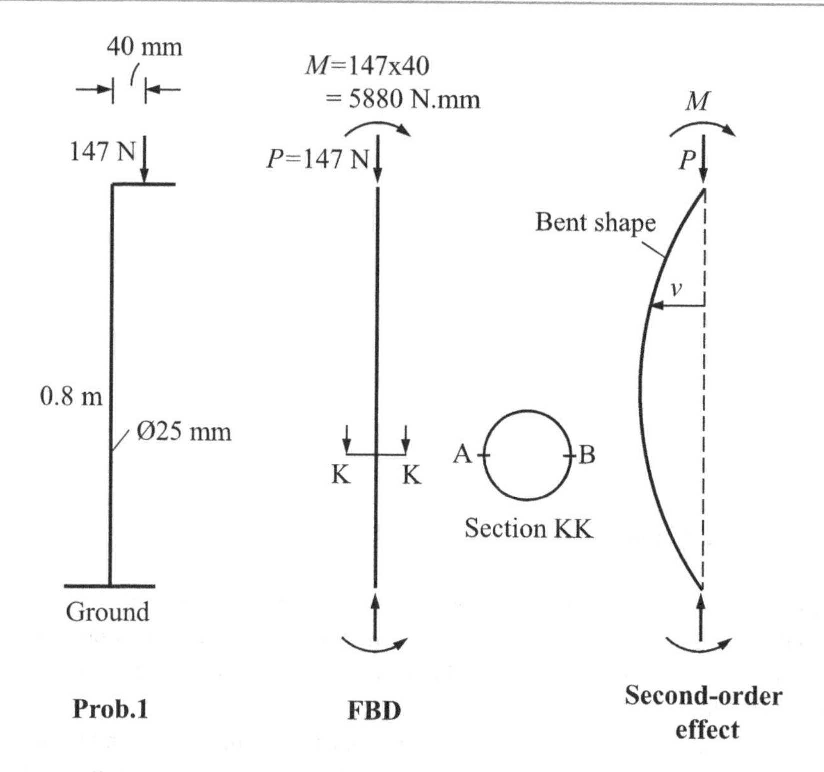

(2) Due to P:

Direct axial stress $= \frac{P}{A} = -\frac{147}{\pi \times 25^2/4} = -0.3$ N/mm^{2}uniform over the cross-section.

(3) Due to M:

Max. bending stress $= \pm\frac{M}{Z} = \pm\frac{5880}{\pi \times 25^3/32} = \pm 3.83$ N/mm^2 at points A, B, respectively.

(4) Thus, all points corresponding to B along the length are critical points with a maximum uniaxial stress = – 4.13 N/mm^2.
(If the material of the stick has different strengths in tension and compression, then location A should also be considered as critical with uniaxial stress = 3.53 N/mm^2. In subsequent problems, the strength is assumed to be the same in tension and compression.)

Problem 2: Find the critical locations/states of stress along the length of the post due to (a) wind pressure of 300 N/m^2 on the signboard; (b) the above wind pressure as well as the weight of the signboard given to be 160 N/m^2.

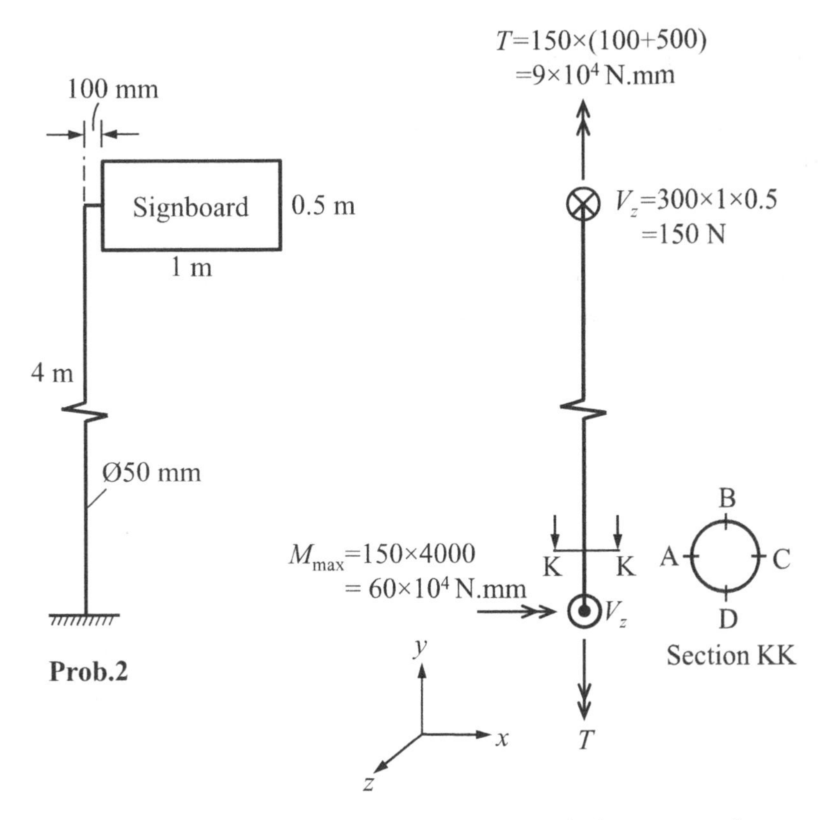

(a) FBD - due to wind pressure alone

Solution steps for (a):

(1) Loads on the post are as shown in FBD.

While V_z and T are invariant along the length, the bending moment due to V_z is maximum at the bottom.

(2) For the circular cross-section, due to V_z:

Max. direct shear stress $= \frac{4}{3}\tau_{av} = \frac{4}{3} \cdot \frac{150}{\pi \times 50^2/4} = 0.10\ \text{N/mm}^2$ at all points along the diameter AC.

Zero shear stress at B, D.

(3) Due to T:

Max. torsional shear stress $= \frac{T}{Z_p} = \frac{9\times10^4}{\pi\times50^3/16} = 3.67\text{N/mm}^2$ at all points of the outer surface.

(4) Due to $M_{\max}$ at the bottom end:

Max. bending stress $= \pm\frac{M_{\max}}{Z} = \pm\frac{60\times10^4}{\pi\times50^3/32} = \pm48.89\ \text{N/mm}^2$ at points D, B, respectively.
No bending stress at A, C.

(5) Transverse and torsional shear stresses are in the same direction at C while they are oppositely directed at A. Thus, C is a critical point under pure shear of 3.77 N/mm^2 as shown.

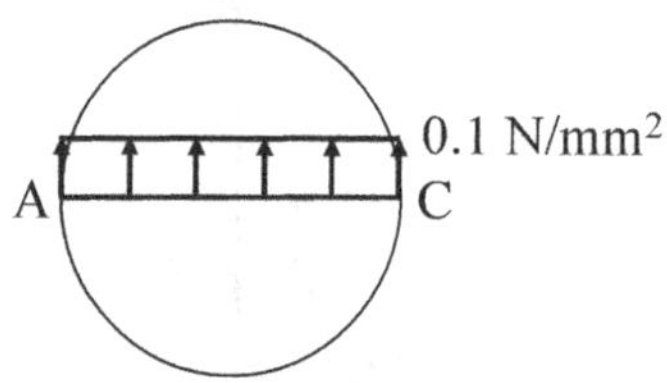

Direct shear stress

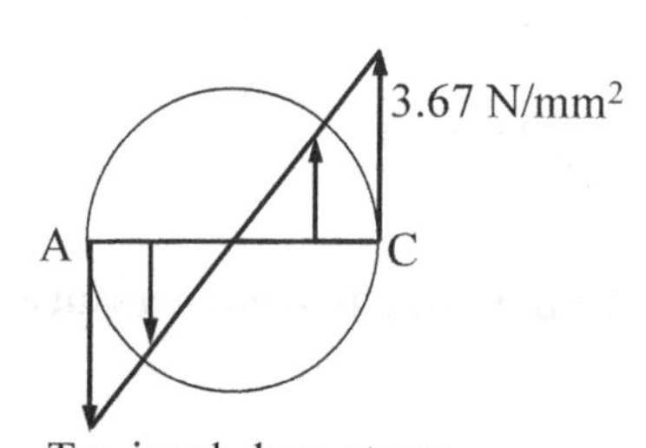

Torsional shear stress

Shear stress variation along AC
(Top sectional view)

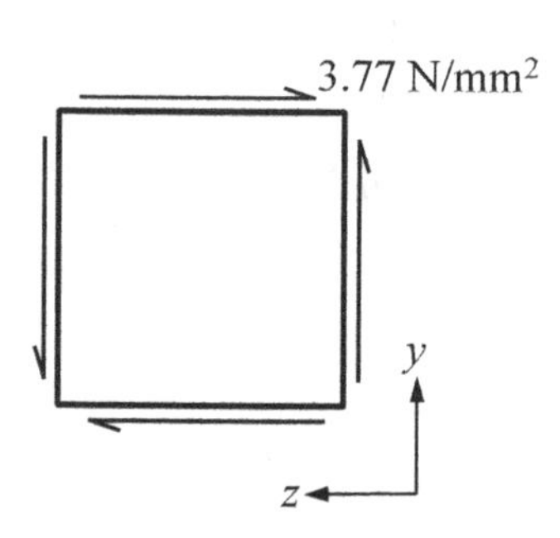

Stresses at C
(Right side view)

(6) B and D are the other critical points with the states of stress as shown.

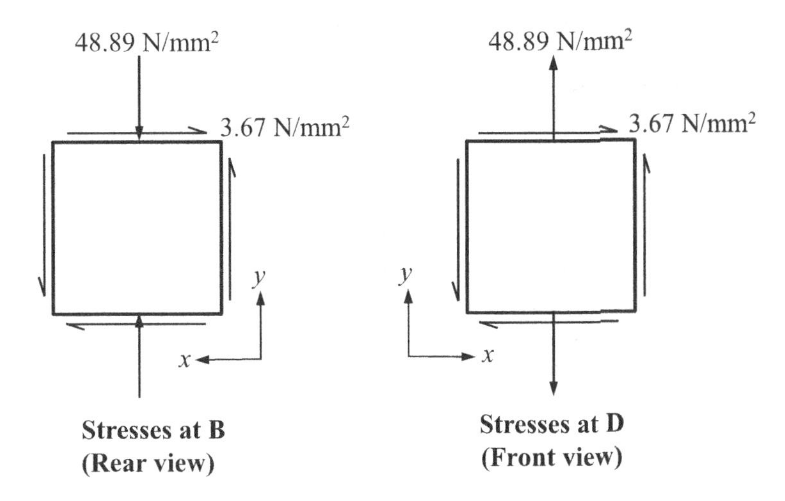

Solution steps for (b):

(1) Loads on the post due to weight of the signboard alone are as shown with the corresponding stresses uniform along the length.

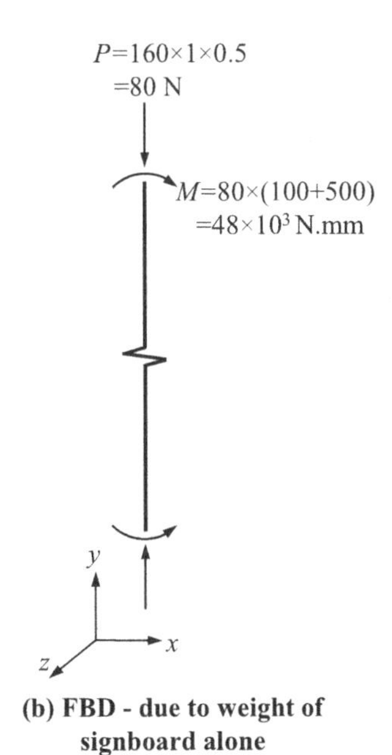

(b) FBD - due to weight of signboard alone

(2) Due to these additional loads alone:
Direct axial stress $= \frac{P}{A}$

$$= -\frac{80}{\frac{\pi\times 50^2}{4}} = -0.04\text{N/mm}^2 \text{uniform over the cross-section.}$$

$$\text{Max. bending stress} = \pm\frac{M}{Z}$$

$$= \pm\frac{48\times 10^3}{\pi\times 50^3/32} = \pm 3.91\text{N/mm}^2 \text{at A, C, respectively.}$$

No bending stress at *B*, *D*.

(3) Thus, due to wind pressure and the weight of the signboard together, *C* and *B* of the bottom cross-section are critical points.
At C : $\sigma_y = -0.04 - 3.91 = -3.95\ \text{N/mm}^2,\ |\tau_{yz}| = 3.77\ \text{N/mm}^2$
At B: $\sigma_y = -0.04 - 48.89 = -48.93\ \text{N/mm}^2,\ |\tau_{xy}| = 3.67\ \text{N/mm}^2$ as shown

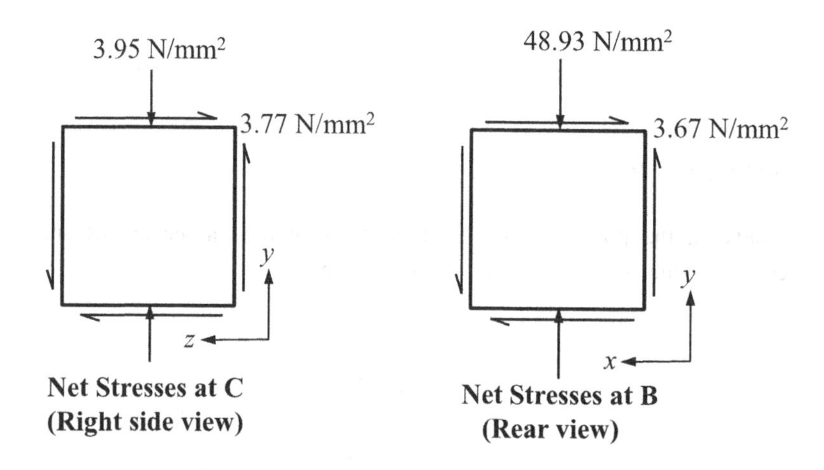

Net Stresses at C (Right side view)

Net Stresses at B (Rear view)

Problem 3: The cylindrical pressure vessel shown may be taken to be simply supported at the ends. Find the states of stress at critical locations if the gross weight of the vessel is 1200 kg.

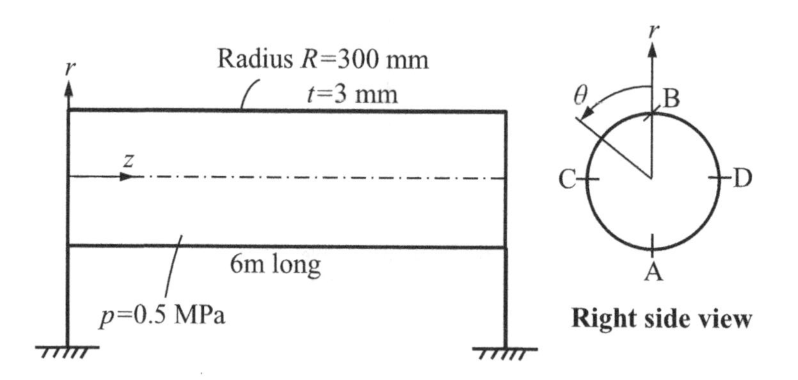

Right side view

Solution steps:

(1) The loads acting on the vessel are internal pressure causing hoop and longitudinal membrane stresses, and its own weight looked upon as uniformly distributed load on a simply supported beam and causing a maximum bending moment of $\frac{qL^2}{8} = \left(\frac{1200\times9.8}{6000}\right)\left(\frac{6000^2}{8}\right) = 8.82 \times 10^6$ N mm at mid-span and a maximum shear force of $\frac{qL}{2} = \frac{1200\times9.8}{2} = 5880$ N at the ends.

(2) Due to internal pressure:
$\sigma_\theta = \frac{pR}{t} = \frac{0.5\times300}{3} = 50$ MPa, $\sigma_z = \frac{pR}{2t} = 25$ MPa, uniform at all points of the thin cylindrical wall.

(3) At mid-span, due to bending moment:
Max. $\sigma_z = \pm\frac{MR}{I} = \pm\frac{MR}{J/2} = \pm\frac{MR}{\pi R^3 t} = \pm\frac{8.82\times10^6}{\pi\times300^2\times3} = \pm10.4$ MPa at points A, B, respectively.

(4) At points C, D of end cross-sections, due to vertical shear force:

$$
\begin{aligned}
|\tau_{z\theta}| &= \frac{VQ}{Ib} = \frac{V\int_{-\pi/2}^{\pi/2}(R\cos\theta\ tRd\theta)}{\pi R^3 t \times 2t} \\
&= \frac{V}{\pi Rt} = \frac{5880}{\pi \times 300 \times 3} = 2.08 \text{ MPa}
\end{aligned}
$$

No shear stress at points A, B of any cross-section.

(5) Hence point A of mid-span cross-section and points C, D of any end cross-section are critical points with states of stress as shown.

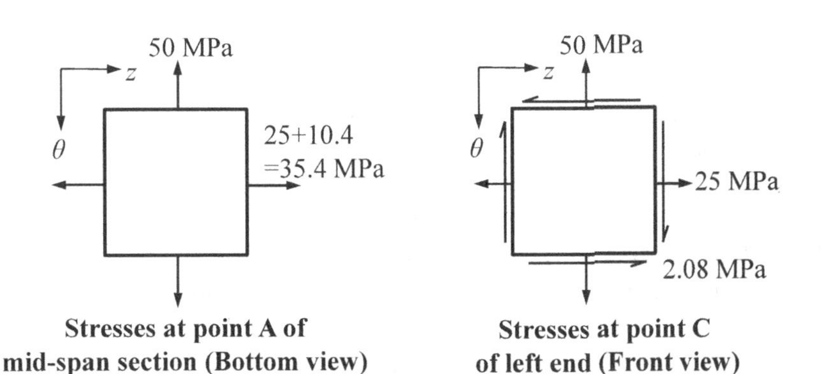

Stresses at point A of mid-span section (Bottom view)

Stresses at point C of left end (Front view)

Problem 4: A stepped power transmission shaft, carrying a driver pulley at E and two driven pulleys at B, C, rotates at constant speed. The shaft is of diameter 34 mm over the length AB and 40 mm over BE. The measured belt tensions are as shown. Assuming the bearings at A, D to act as simple supports, and taking transverse shear stresses as negligible, find the critical points/states of stress in the shaft.

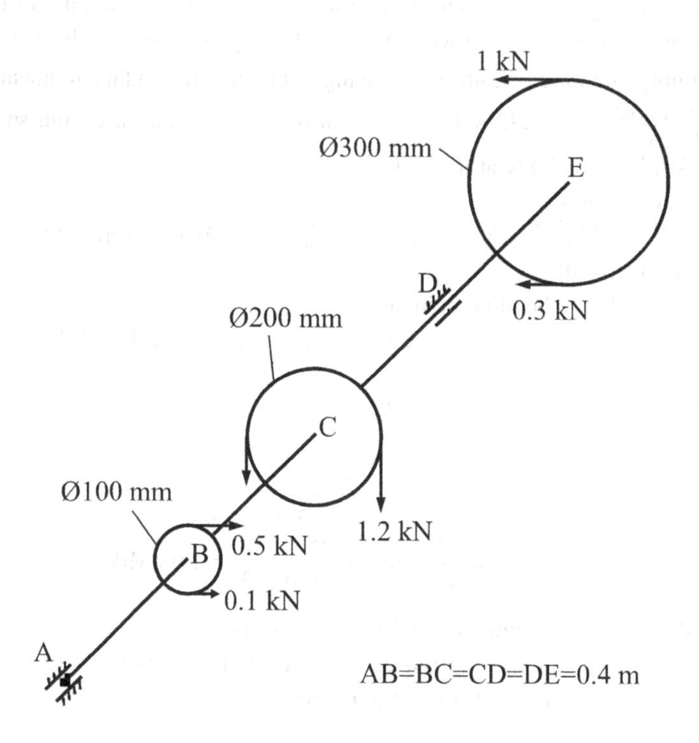

Solution steps:

(1) Torque at $E = (1-0.3)150 = 105$ N m ↺
Torque at $B = (0.5-0.1)50 = 20$ N m ↻
Hence, torque at $C = 105-20 = 85$ N m ↻
Thus, if the unknown belt tension is x at C, then
$(1.2-x)100 = 85 \Rightarrow x = 0.35$ kN
(2) Torque variation along the length of the shaft is as shown.
(3) The vertical and horizontal loading are considered separately as shown, and corresponding BMDs drawn. Net bending moments at B, C, D are obtained by vector addition as $\sqrt{(206.67^2+333.33^2)}$ =392.2 N m, $\sqrt{(413.33^2+426.67^2)}$ =594.04 N m, and 520 N m, respectively. (Corresponding inclinations of the neutral axis and the extreme fiber locations with maximum bending stress may easily be identified, but not done here).

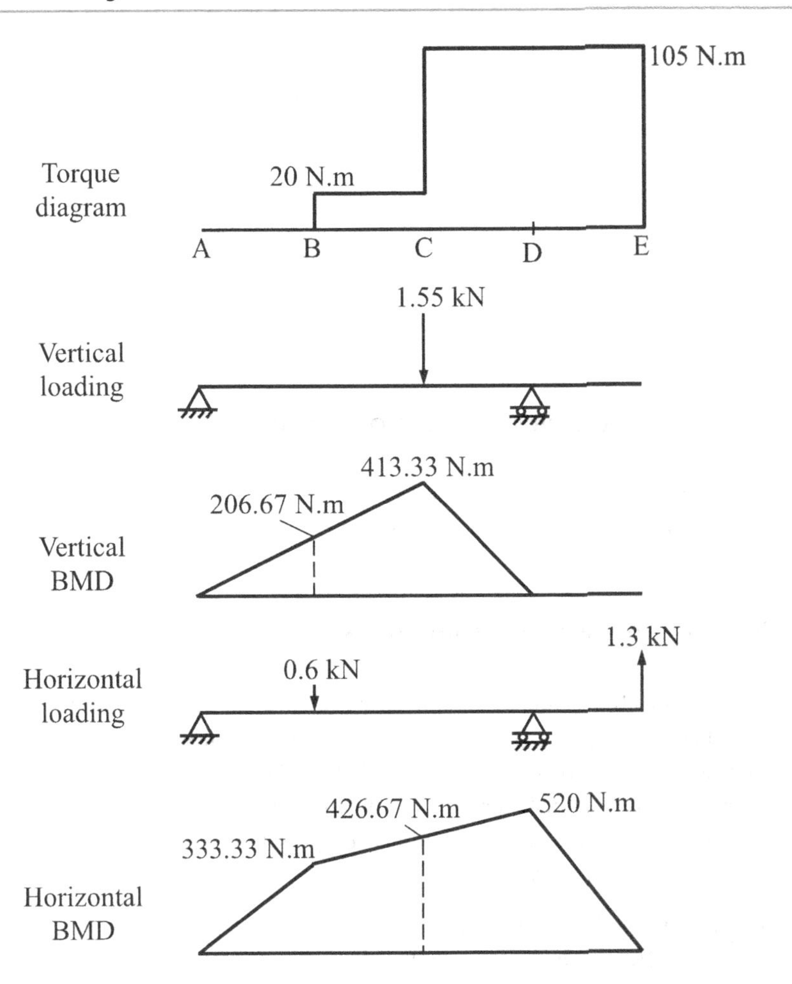

(4) Over length *AB* of diameter 34 mm, *B* is the critical section with $M = 392.2$ N m and $T = 0$.

Corresponding $\sigma_{\max} = \pm\frac{M}{Z} = \pm\frac{392.2\times10^3}{\pi\times34^3/32} = \pm101.6$MPa. This is a state of simple uniaxial stress.

(5) Over length BE of diameter 40 mm, the section just to the right of *C* is most severely loaded with $M = 594.04$ N.m and $T = 105$ N m. These yield $\sigma_{\max} = \pm94.5$ MPa, $\tau_{\max} = 8.4$ MPa at the critical extreme fiber locations farthest from the neutral axis. The element on the tension side alone is shown.

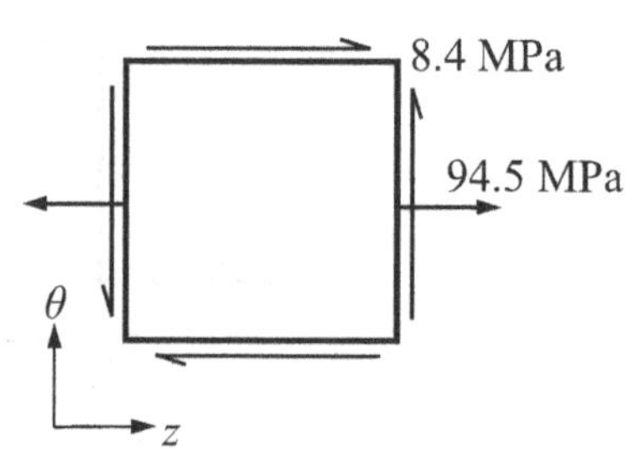

8.2 Stress and Strain Transformation

This may be carried out using several techniques as explained below. All the problems of this section pertain to a body in plane stress.

8.2.1 Use of Transformation Rules

The use of transformation rules and formulae for principal stresses and strains and their planes of action, as put forth in Sects. 1.8–1.11, is illustrated in the following problems. In simple cases as in Problem 5, a direct fundamental approach based on equilibrium considerations (or kinematic considerations to obtain strains) for suitable infinitesimal elements may be used. (In fact, the transformation rules presented in Chap. 1 are themselves derived using such a methodology.)

Problem 5: A wooden rectangular bar of cross-sectional area A is obtained by bonding two pieces along QRS as shown. If the allowable stress for the adhesive bond is σ_0 in tension and $2\sigma_0$ in shear, find the allowable tensile load P for the bar. Solve this by considering suitable infinitesimal elements and invoking equilibrium considerations directly.

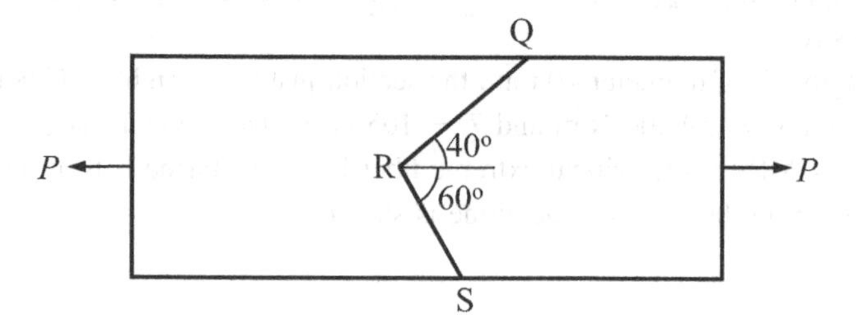

Solution steps:

(1) Consider an infinitesimal triangular element at any point on the face QR as shown; let its hypotenuse as well as the thickness perpendicular to the page be unity. The stress on the vertical face is just the normal stress of *P/A* while the horizontal face is stress-free.

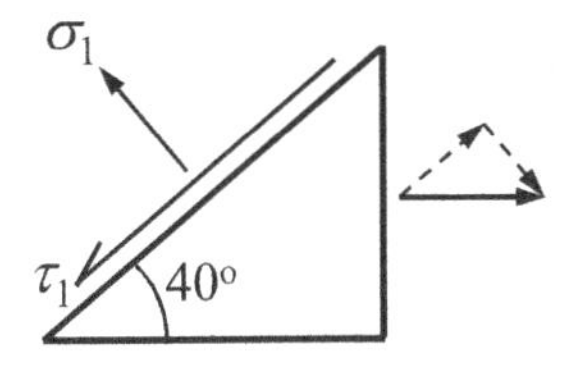

If the stresses on the slant face are taken as σ_1 and τ_1 as shown, then, for equilibrium of forces in the directions of these stresses,

$$\sigma_1 \times 1 \times 1 = \left(\frac{P}{A}\right) \times \left(1 \sin 40^\circ\right) \sin 40^\circ \Rightarrow \sigma_1 = 0.41\frac{P}{A}$$

$$\tau_1 \times 1 \times 1 = \left(\frac{P}{A}\right) \times \left(1 \sin 40^\circ\right) \cos 40^\circ \Rightarrow \tau_1 = 0.49\frac{P}{A}$$

(2) Similarly, for a triangular element at any point on the face RS as shown,

$$\sigma_2 = \left(\frac{P}{A}\right) \sin^2 60^\circ = 0.75\frac{P}{A}$$

$$\tau_2 = \left(\frac{P}{A}\right) \sin 60^\circ \cos 60^\circ = 0.43\frac{P}{A}$$

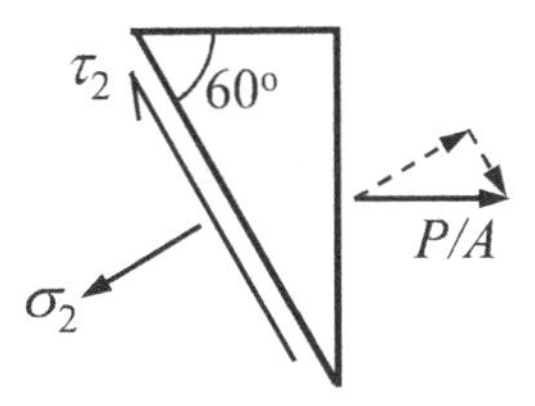

(3) Max. normal stress on adhesive $= 0.75\frac{P}{A} = \sigma_o \Rightarrow P = 1.33\sigma_\alpha A$
Max. shear stress on adhesive $= 0.49\frac{P}{A} = 2\sigma_0 \Rightarrow P = 4.08\sigma_\alpha A.$
Hence, $P_{\text{all}} = 1.33\sigma_o A.$

Problem 6: The state of stress at a point is given by . $\sigma_x = 80\text{MPa}$, $\sigma_y = 60$ MPa, $\tau_{xy} = 40\text{MPa}$ $\tau_{xy} = 40\text{MPa}$ Using the equations of Sects. 1.8–1.11, find (a) the state of stress with respect to x'–y' axes obtained by rotating x–y axes by 30° anticlockwise; (b) the principal stresses and maximum in-plane shear stress, showing them on suitably oriented elements.

Solution steps:

(1) The direction cosines of x'–y' axes are:

$$a_{x'x} = \cos 30^{\circ} = \frac{\sqrt{3}}{2}$$

$$a_{x'y} = \cos 60^{\circ} = \frac{1}{2}$$

$$a_{y'x} = \cos 120^{\circ} = -\frac{1}{2}$$

$$a_{y'y} = \cos 30^{\circ} = \frac{\sqrt{3}}{2}$$

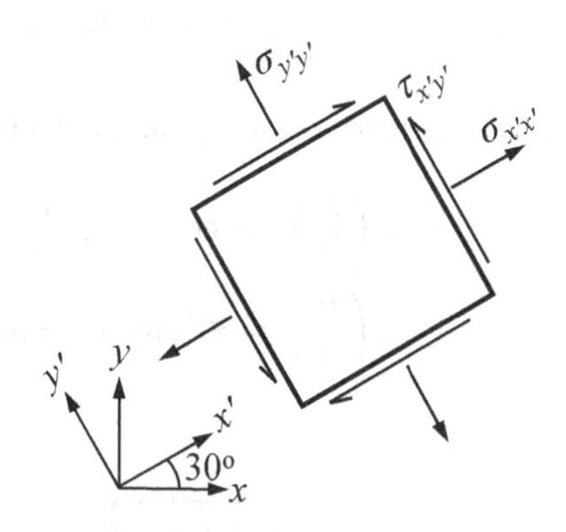

(2) $\sigma_{x'x'} = a^2_{x'x^2}\sigma_{xx} + a_{x'x}a_{x'y}\tau_{xy} + a_{x'y}a_{x'x}\tau_{yx} + a^2_{x'y}\sigma_{yy} = 109.64$ MPa
$\tau_{x'y'} = a_{x'x}a_{y'x}\sigma_{xx} + a_{x'x}a_{y'y'}\tau_{xy} + a_{x'y}a_{y'x}\tau_{yx} + a_{x'y}a_{y'y}\sigma_{yy} = 11.34\text{MPa}$
$\sigma_{y'y'} = a^2_{y'x}\sigma_{xx} + a_{y'x}a_{y'y}\tau_{xy} + a_{y'y}a_{y'x}\tau_{yx} + a^2_{y'y}\sigma_{yy} = 30.36$ MPa

(3) $$\sigma_{P_1}, \sigma_{P_2} = \frac{\sigma_x + \sigma_y}{2} \pm \sqrt{\left(\frac{\sigma_x - \sigma_y}{2}\right)^2 + \tau^2_{xy}} = 70 \pm \sqrt{10^2 + 40^2} = 111.23, 28.77 \text{ MPa}$$

(4) $\left|\tau_{\text{max in - plane}}\right| = \left|\frac{\sigma_{P_1} - \sigma_{P_2}}{2}\right| = 41.23$ MPa

(5) Principal plane orientations are obtained from

$$\tan 2\theta = \frac{2\tau_{xy}}{(\sigma_x - \sigma_y)} = 4 \Rightarrow \theta = 37.98^\circ, 90 + 37.98 = 127.98^\circ$$

where θ is the angle made by the normal to the plane with respect to the x-axis and taken as positive when measured anticlockwise (see Fig. 1.4).

To assign σ_{P1} and σ_{P2} correctly to the two principal planes, it is necessary to substitute one of the above values of θ in the stress transformation equation for $\sigma_{x'x'}$. This procedure results in the following rule:

"The major (algebraically larger) principal stress σ_{P1} acts on the principal plane which is closer to (i.e. less than 45° away from) the plane of the algebraically larger normal stress in the given x-y coordinate system."

Thus, here, the plane of σ_{P1} is closer to the vertical plane as shown.

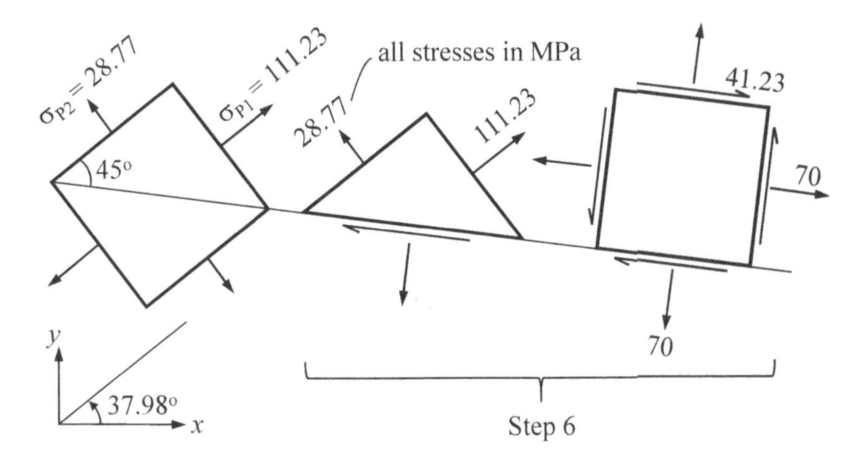

(6) The planes corresponding to $|\tau_{\text{max in-plane}}|$ are 45° away from the principal planes. The direction of the shear stress on one of them is fixed up by considering the equilibrium of an isosceles triangular element as shown, and that on the orthogonal plane using the principle of complementary shear. Note that the state of stress with reference to these max. shear planes must include the equal biaxial normal stresses of $\left(\frac{\sigma_x+\sigma_y}{2}\right) = 70\text{MPa}$ as shown.

Problem 7: A small square of side 4 mm is drawn on a thin sheet. When the sheet is subjected to edge loading, the square deforms into a parallelogram with sides of 4.004 mm and 3.998 mm, and with one of the diagonals as 5.67 mm. Find the interior angles of the parallelogram.

Solution steps:

(1) Let the undeformed square be as shown with x, y axes aligned with its sides and x', y' axes with its diagonals. Let AB, CD be the elongated sides and AD, BC be the compressed ones, and let AC be the diagonal with final length equal to 5.67 mm.

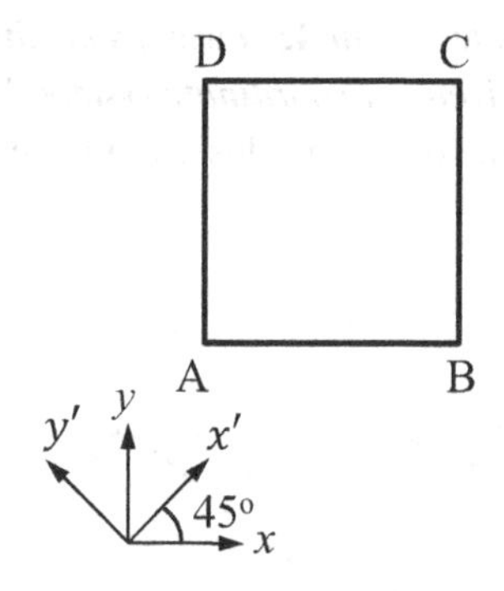

Then, one has

$$\varepsilon_x = \frac{4.004 - 4}{4} = 0.001;$$

$$\varepsilon_y = \frac{3.998 - 4}{4} = -0.0005;$$

$$\varepsilon_{x'} = \frac{5.67 - 4\sqrt{2}}{4\sqrt{2}} = 2.324 \times 10^{-3}$$

(2) $\varepsilon_{x'x'} = a^2_{x'x^2}\varepsilon_{xx} + a_{x'x}a_{x'y'}\varepsilon_{xy} + a_{x'y}a_{x'x}\varepsilon_{yx} + a^2_{x'y^2}\varepsilon_{yy}$

$$\Rightarrow 2.324 \times 10^{-3} = 0.001 \cos^2 45^\circ + 2\varepsilon_{xy} \cos 45^\circ \sin 45^\circ - 0.0005 \sin^2 45^\circ$$

yielding $\varepsilon_{xy} = 2.074 \times 10^{-3}$,
i.e.$\gamma_{xy} = 2\varepsilon_{xy} = 4.148 \times 10^{-3}$.

(3) The required interior angles are $\left(\frac{\pi}{2} \pm \gamma_{xy}\right)$ rad, i.e. 90.24° at corners B, D and 89.76° at corners A, C as shown, corresponding to positive γ_{xy}.

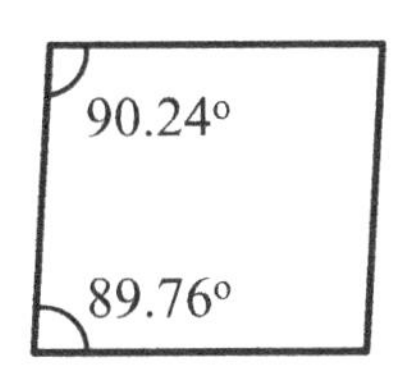

Problem 8: At a point, σ_x and τ_{xy} are known to be 100 MPa and 40 MPa, respectively. Find the allowable range of σ_y if the tensile or compressive $\sigma_{\text{allowable}}$ for the material is 200 MPa.

Solution steps:

(1) One of the extreme values of σ_y would correspond to $\sigma_{P1} = -200$ MPa, and the other to $\sigma_{P2} = -200$ MPa.

(2) Thus, $\frac{100+\sigma_y}{2} + \sqrt{\left(\frac{100-\sigma_y}{2}\right)^2 + 40^2} = 200,$
 yielding $\sigma_y = 184$MPa (and corresponding $\sigma_{P2} = 84$MPa).

(3) Similarly, $\frac{100+\sigma_y}{2} - \sqrt{\left(\frac{100-\sigma_y}{2}\right)^2 + 40^2} = -200$
 yields $\sigma_y = -194.67$MPa (and corresponding $\sigma_{P1} = 105.33$MPa).

(4) Thus, the allowable range of σ_y is -194.67 MPa to $+184$ MPa.

8.2.2 Mohr's Circle

This is a semi-graphical aid for carrying out stress or strain transformation. It is a *free-hand sketch*, drawn approximately to scale, with the normal stress (or normal strain) as the abscissa and the shear stress (or tensorial shear strain) as the ordinate. When plotted thus, the locus of points corresponding to states of stress or strain with reference to rotated coordinate axes is a circle.

It should be noted that Mohr's circle is simply a graphical representation of the transformation rules put forth in Chap. 1 and is hence fundamentally the same. Thus, its use vis-à-vis that of the analytical approach is essentially one of personal preference, or slightly greater convenience because of the visual representation and that of not having to remember the transformation rules.

Given the two-dimensional state of stress or strain in x–y coordinates, the corresponding Mohr's circle may be drawn as explained below; each step is illustrated here with reference to Prob.6 (i.e. $\sigma_x = 80$ MPa, $\sigma_y = 60$ MPa, $\tau_{xy} = 40$ MPa), with corresponding remarks in parentheses.

Step 1: Show the stress or strain components on an elemental square with correct orientations as per the usual sign convention.

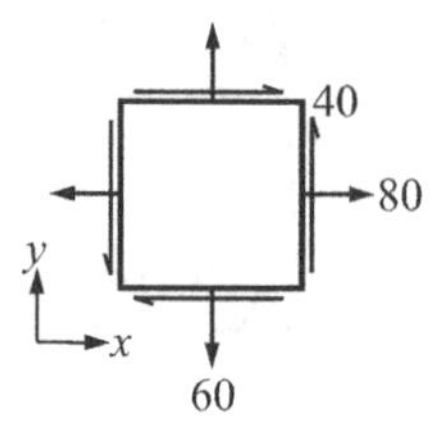

Step 2: Considering the shearing action on opposite faces of the element, designate the shear on vertical and horizontal planes as "clockwise shear" or "anticlockwise shear".

(Here, anticlockwise shear on vertical plane and clockwise shear on horizontal plane.)

Step 3: Identify coordinates for the points corresponding to the vertical and horizontal planes (designated here as* V *and* H *, respectively); the first coordinate is the normal stress or strain (positive if tensile, negative if compressive) and the second coordinate is the shear stress or tensorial shear strain (clockwise or anticlockwise).

(V(80, 40↺), H(60, 40↻) here.)

Step 4 : Plot the points* V *and* H *with normal stresses or strains (σ or ε) on the abscissa (tensile stresses or strains to the right and compressive ones to the left), and the shear stresses or tensorial shear strains (τ or $\frac{\gamma}{2}$) on the ordinate (clockwise shear upwards and anticlockwise downwards). Join* V *and* H *to obtain a diameter and complete the circle.

Step 5: The circle may be employed to extract further information, using <u>purely geometric considerations</u> and the following features:

- ***Each point of the circle represents an inclined plane and its coordinates are the corresponding normal and shear stresses (or normal and tensorial shear strains)***
- ***Any point* K *on this circle, located at 2 θ anticlockwise with respect to* V *, corresponds to the inclined plane normal to x' oriented at θ anticlockwise with respect to x as shown. Thus, the diametrically opposite point* L *would correspond to the plane normal to y' .***
- ***End-points P_1, P_2 of the horizontal diameter correspond to the principal planes—with maximum and minimum σ, respectively, and zero τ.***
- ***End-points T_1, T_2 of the vertical diameter correspond to planes of maximum in-plane shear—with $\sigma = \frac{(\sigma_x+\sigma_y)}{2}$.***

(Thus, one gets the following:

- Center C of the circle at $\left(\frac{80+60}{2}, 0\right)$ or (70, 0)
- Radius $= \mathrm{CV} = \sqrt{\mathrm{CV}_1^2 + \mathrm{VV}_1^2} = \sqrt{(80-70)^2 + 40^2} = 41.23$
- P_1(70 + 41.23,0) or (111.23,0), P_2(70− 41.23,0) or (28.77,0)

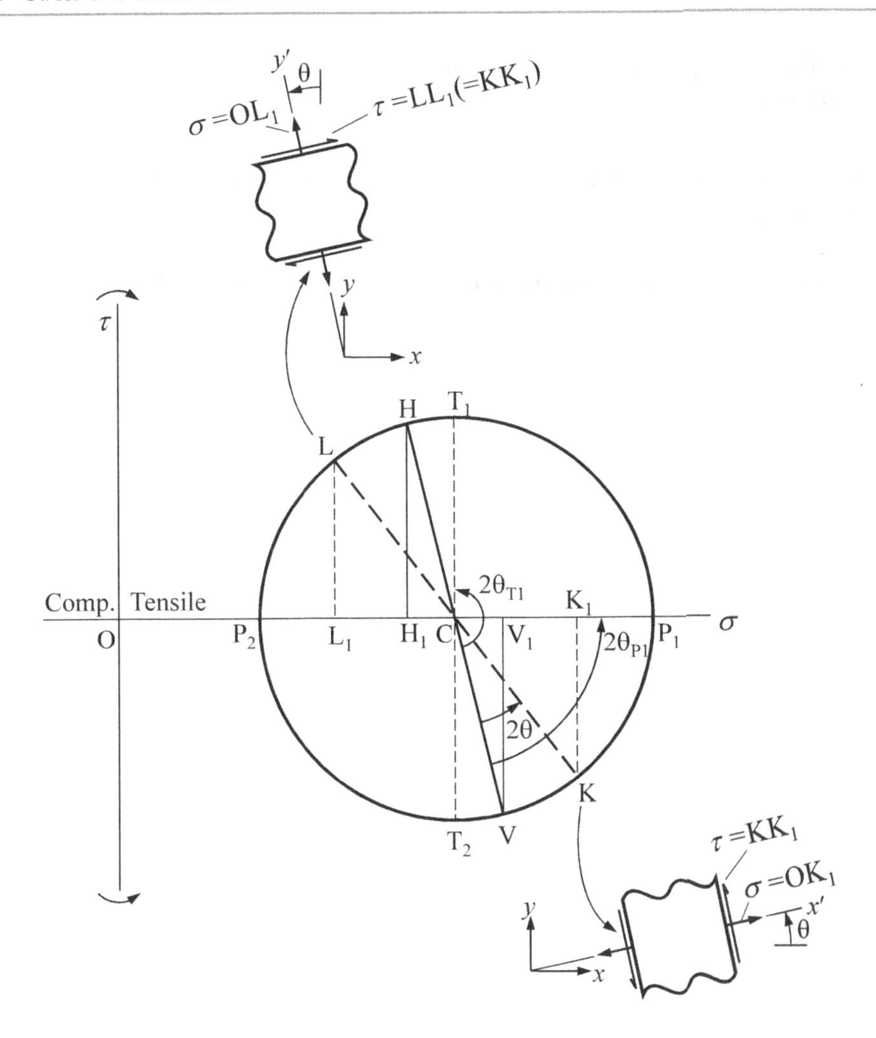

- $T_1(70, 41.23↻)$, $T_2(70, 41.23↺)$
- $\tan 2\theta_{P1} = \frac{VV_1}{CV_1} = \frac{40}{80-70} \Rightarrow \theta_{P1} = 37.98^{\circ}$
- $\theta_{T1} = 45^{\circ} + 37.98^{\circ} = 82.98^{\circ}$
- For $2\theta = 2 \times 30^{\circ} = 60^{\circ}$, K is at

  $(\mathrm{OC} + \mathrm{CK}\cos(2\theta_{P1} - 2\theta), \mathrm{CK}\sin(2\theta_{P1} - 2\theta)$

  i.e. $(70 + 41.23\cos(75.96^{\circ} - 60^{\circ}), 41.23\sin(75.96^{\circ} - 60^{\circ})↺)$
  i.e. $(70 + 39.64 = 109.64, 11.34↺)$
  Hence, L is at $(70 - 39.64 = 30.36, 11.34↻)$

- Suitably oriented elements showing σ_{P1}, σ_{P2} and $\tau_{\text{max in-plane}}$ are exactly as in Problem 6).

Mohr's circle is shown below for several elementary cases of loading, with due observations therefrom.

(i) Long bar in tension or compression – Uniaxial stress with $\sigma_x \neq 0$, $\sigma_y = \tau_{xy} = 0$

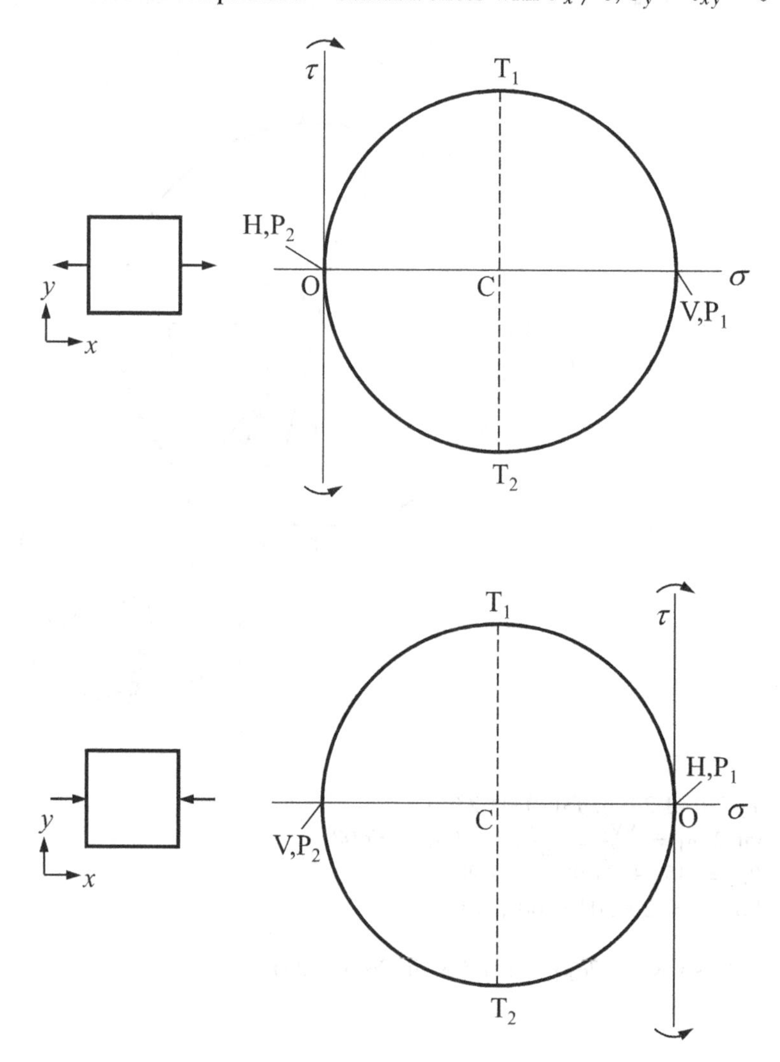

Observation: $\left|\tau_{\text{max in-pl.}}\right| = \frac{|\sigma_x|}{2}$

(ii) Thin-walled cylindrical pressure vessel under internal or external pressure—Biaxial stresses with $\sigma_y = 2\sigma_x$, $\tau_{xy} = 0$ (x , y along axial and circumferential directions, respectively)

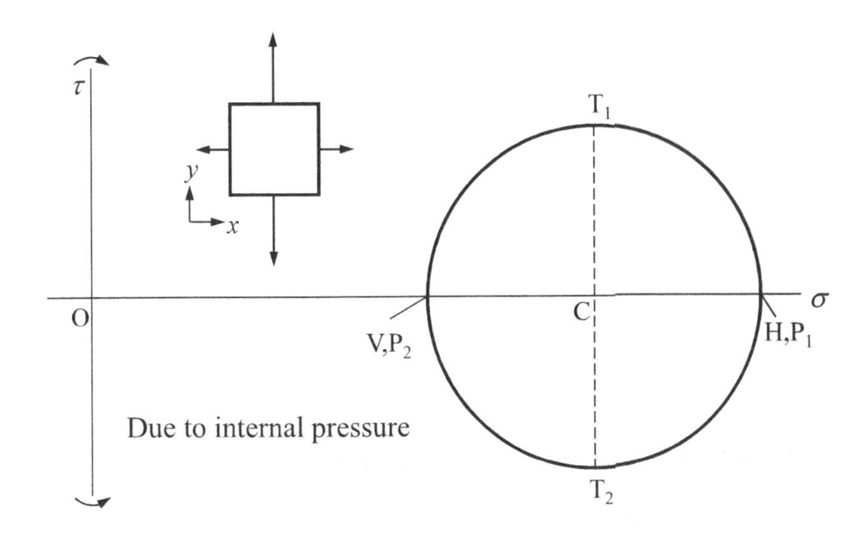

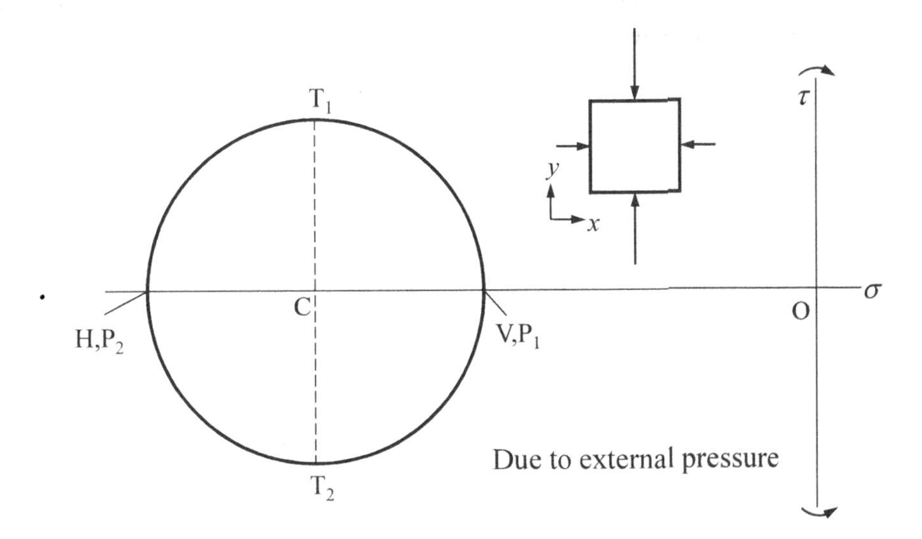

Observation: $\left|\tau_{\text{max in-pl.}}\right| = \frac{|\sigma_x - \sigma_y|}{2}$

(iii) Spherical pressure vessel under internal or external pressure—Equal biaxial stresses: $\sigma_x = \sigma_y,\ \tau_{xy} = 0$

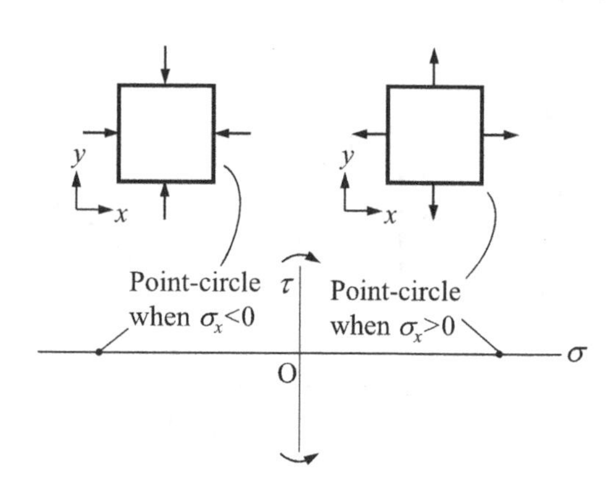

Observations :

(a) Isotropic biaxial state—same for all rotatedx'–y' axes.

(b) $\left|\tau_{\text{max in-pl.}}\right| = 0$

(iv) Circular shaft under torsion—Pure shear with $\tau_{xy} \neq 0,\ \sigma_x = \sigma_y = 0$ (x , y along axial and circumferential directions, respectively)

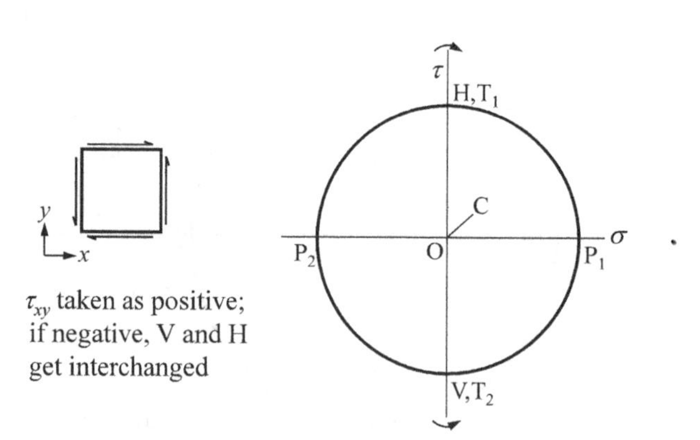

Observation: $\sigma_{P1} = -\sigma_{P2} = \left|\tau_{xy}\right|$

(v) Equal and unlike biaxial stresses with $\sigma_x = -\sigma_y$, $\tau_{xy} = 0$

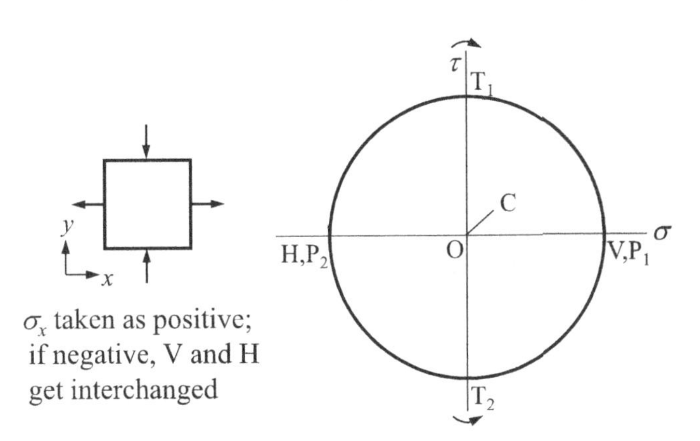

Observations:

(a) $|\tau_{\text{max in-pl.}}| = \sigma_x$

(b) This is a case of pure shear with respect to x'–y' axes 45° away from x–y (i.e. corresponding to T_1 and T_2).

(c) Mohr's circle same as for Case (iv) except for locations of V and H.

(vi) Equal and unlike biaxial stresses along with shear: $\sigma_x = -\sigma_y$, $\tau_{xy} \neq 0$, $\sigma_x > 0$

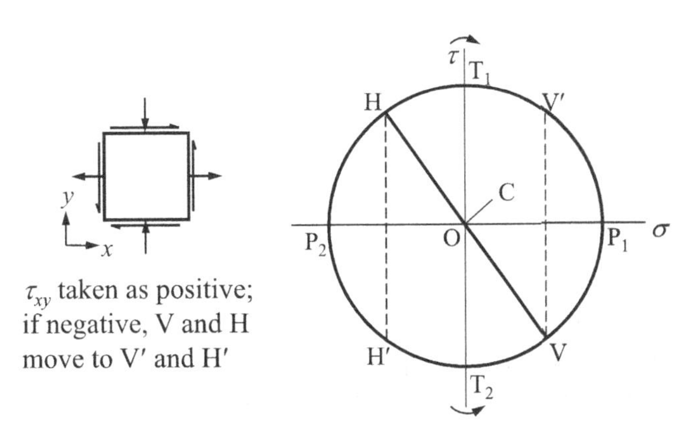

Observations:

(a) This is also a case of pure shear with respect to rotated axes corresponding to T_1 and T_2.

(b) $|\tau_{\text{max in-pl.}}| = \sqrt{\sigma_x^2 + \tau_{xy}^2}$.

(vii) Combined bending and torsion of a circular shaft: $\sigma_x \neq 0$, $\tau_{xy} \neq 0$, $\sigma_y = 0$ at the two critical points farthest from the neutral axis

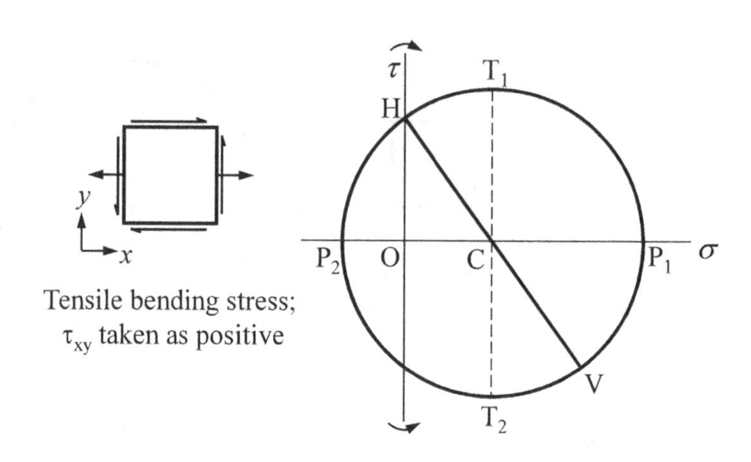

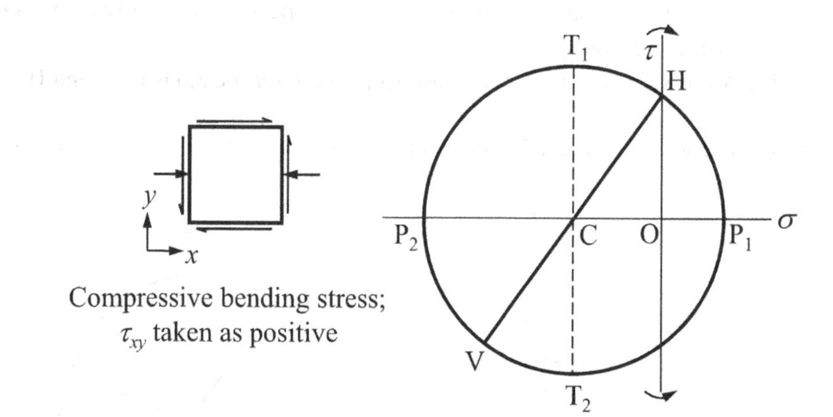

Observations:

(a) Maximum tensile and compressive stresses in the shaft occur at the two extreme points, corresponding to σ_{P1} on the tension side and σ_{P2} on the compression side, respectively; both of them are of the same magnitude as given by $\frac{|\sigma_x|}{2} + \sqrt{\left(\frac{\sigma_x}{2}\right)^2 + \tau_{xy}}$.

(b) $\left|\tau_{\text{max in-pl.}}\right| = \sqrt{\left(\frac{\sigma_x}{2}\right)^2 + \tau_{xy}^2}$ at any of these extreme points.

The following problems further illustrate the use of Mohr's circle.

Problem 9: Sketch the Mohr's circle corresponding to Problem 7, and using it, find the orientation of a square which, if drawn at the same location as earlier, would have deformed into (i) a rectangle and not a parallelogram; (ii) a rhombus. For (i), find the aspect ratio of the rectangle, and for (ii), the interior angles of the rhombus.

Solution steps:

(1) Positive ε_{xy} corresponds to anticlockwise shear on vertical plane and clockwise shear on horizontal plane as shown.

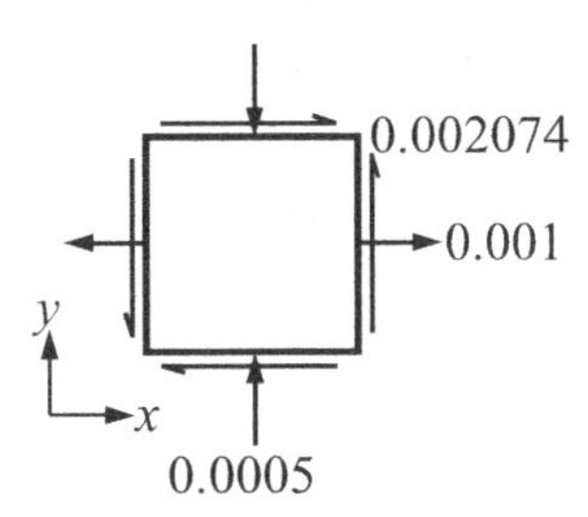

Tensorial strains

(2) Hence, $V(0.001, 2.074 \times 10^{-3}$ ↺) and $H(-0.0005, 2.074 \times 10^{-3}$ ↻), and Mohr's circle as shown below.

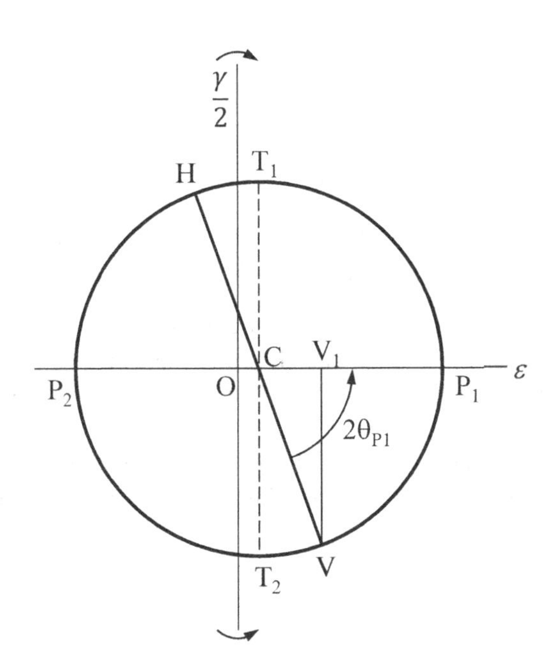

(3) A deformed rectangular shape implies absence of shear distortion and hence corresponds to an original square oriented along the principal axes; similarly, a deformed rhombic shape implies $\varepsilon_{x'} = \varepsilon_{y'}$ and hence corresponds to points T_1 and T_2.

(4) C is at $\left(\frac{0.001-0.0005}{2}, 0\right)$ or (0.00025, 0)

$$\tan 2\theta_{P1} = \frac{\mathrm{VV_1}}{\mathrm{CV_1}} = \frac{2.074 \times 10^{-3}}{0.001 - 0.00025} \Rightarrow \theta_{P1} = 35.06^\circ$$

$$\theta_{T2} = 35.06^\circ - 45^\circ = -9.94^\circ$$

$$\mathrm{CV} = \sqrt{\mathrm{CV_1^2} + \mathrm{VV_1^2}} = 2.205 \times 10^{-3}$$

$$\mathrm{P_1(OC + CV, 0)} \text{ or } \left(2.455 \times 10^{-3}, 0\right)$$

$$\mathrm{P_2(OC - CV, 0)} \text{ or } \left(-1.955 \times 10^{-3}, 0\right)$$

$|\gamma_{max}|$ corresponding to $T_1, T_2 = 2 \times 2.205 \times 10^{-3}$ rad or 0.253°.

(5) Thus, orientation of the original square should be 35.06° ↺ for (i), and 9.94° ↻ for (ii), as shown.

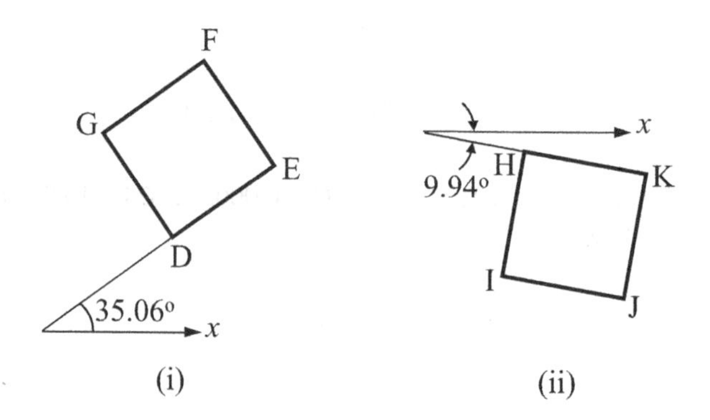

(6) For (i), aspect ratio of the rectangle is given by

$$\frac{\text{Deformed length of DE}}{\text{Deformed length of DG}} = \frac{4(1 + \varepsilon_{P1})}{4(1 + \varepsilon_{P2})} = 1.0044$$

(7) For (ii), the shearing action is clockwise on the planes HK and IJ (corresponding to T_1) and anticlockwise on HI and JK (corresponding to T_2). Thus, interior angles of the resulting rhombus would bc 90°−0.253° = 89.747° (at I, K) and 90.253° (at H, J).

Problem 10: Given that $\sigma_{P1} = 50$ MPa and that there are two normal-stress-free planes 45° away from each other, find the magnitude of the shear stress on these planes and σ_{P2}.

Solution steps:

(1) In Mohr's circle, the normal-stress-free planes correspond to intersection points with the vertical τ-axis. Two possibilities exist as shown, with $\angle ACB = 2(45°)$.

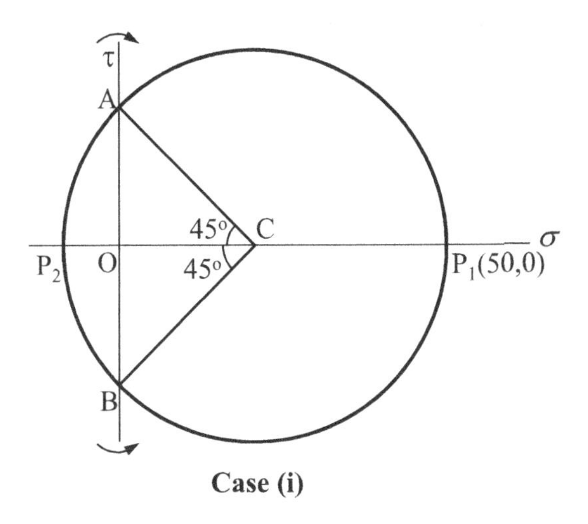

Case (i)

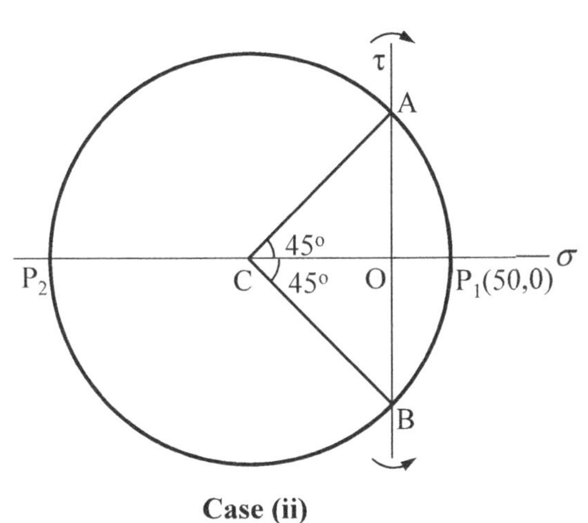

Case (ii)
(not to the same scale as (i))

(2) For Case (i),

$OP_1 = 50 = OC + CP_1 = CA \cos 45° + CP_1 = CA(1 + \cos 45°)$
yielding $CA = 29.29$.
$OP_2 = CP_2 - OC = CA(1 - \cos 45°) = 8.58$
$OA = OB = CA \sin 45° = 20.71$

Thus, $|\tau| = 20.71$ MPa on the two planes corresponding to points A and B, and $\sigma_{P2} = -8.58$ MPa.

(3) Similarly, for Case (ii), one gets
$|\tau| = 120.71$ MPa, $\sigma_{P2} = -291.42$ MPa

(4) The states of stress are as shown.

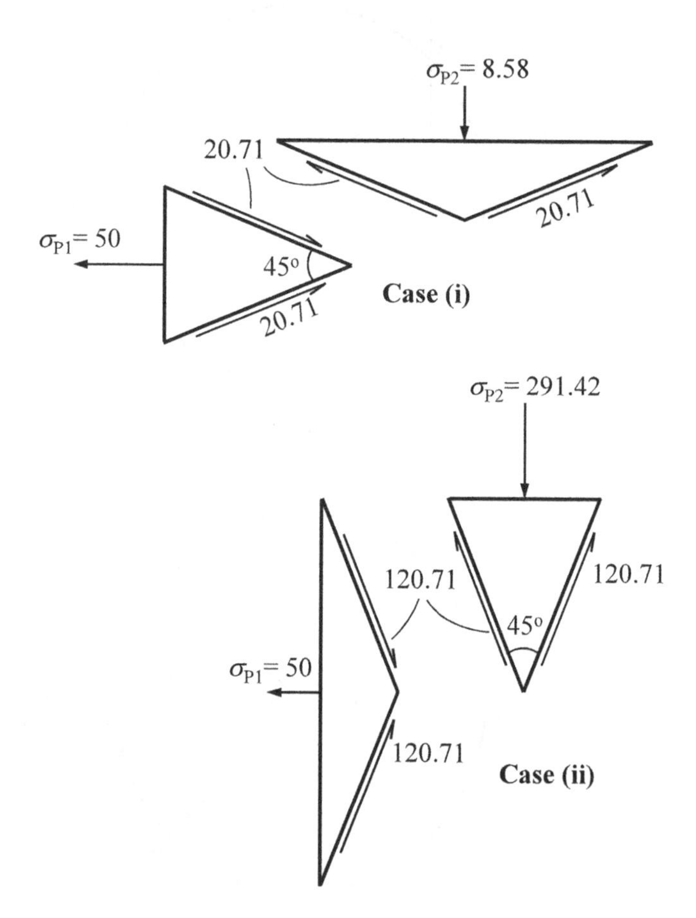

Problem 11: At a point, the normal and shear stresses on two planes are given as $\sigma = 100$ MPa, $|\tau| = 50$ MPa on plane A, and $\sigma = 70$ MPa, $|\tau| = 100$ MPa on plane *B*. Find (a) the principal stresses; (b) the angle between planes *A* and *B*.

Solution steps:

(1) The two planes correspond to two points on Mohr's circle—A_1 or A_2 at (100, 50 ↻ or ↺), and B_1 or B_2 at (70, 100 ↻ or ↺) as shown.

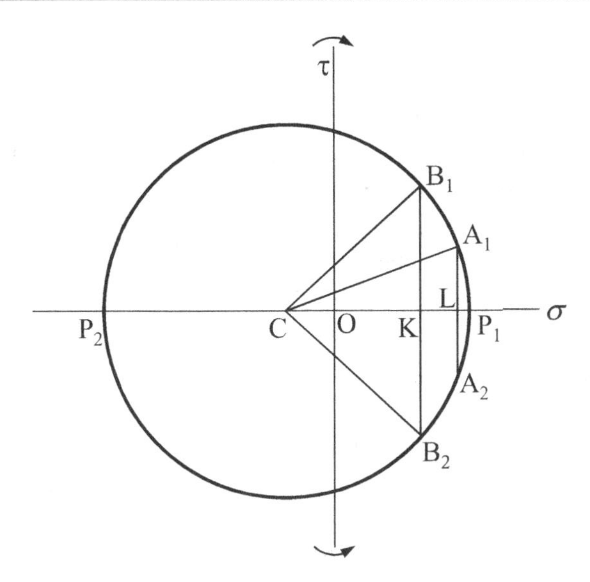

(2) If the centre C is at $(x, 0)$, then C is equidistant from these points.
Hence, $(x - 100)^2 + 50^2 = (x - 70)^2 + 100^2 \Rightarrow x = -40.$

(3) $CB_1 = \sqrt{110^2 + 100^2} = 148.66$

$$\text{OP}_1 = 148.66 - 40 = 108.66, \text{OP}_2 = 148.66 + 40 = 188.66$$

Thus, $\sigma_{P1} = 108.66$ MPa, $\sigma_{P2} = -188.66$ MPa

(4) $\angle B_1CA_1 = \angle B_1CP_1 - \angle A_1CP_1$

$$= \tan^{-1}\frac{\text{B}_1\text{K}}{\text{CK}} - \tan^{-1}\frac{\text{A}_1\text{L}}{\text{CL}} = \tan^{-1}\frac{100}{110} - \tan^{-1}\frac{50}{140} = 22.62^\circ$$

Similarly, $\angle B_2CA_1 = \angle B_2CP_1 + \angle A_1CP_1 = 61.92^\circ$

(5) Thus, the angle between planes A and $B = 11.31^\circ$ or 30.96°.

8.3 Use of Failure Criteria

The failure criteria of Sects. 1.14 and 1.15 are employed here to decide the minimum dimensions required for a structural member or to find the factor of safety for a member of given dimensions. The factor of safety (FoS) is the ratio of the failure strength of the material to the maximum stress occurring in the structure.

Problem 12: The maximum shear stress in a prismatic circular shaft subjected to pure torsion is given as 100 MPa. Find FoS if (i) the material is brittle and $\sigma_{\text{ult. tensile}} = \sigma_{\text{ult. compressive}} = 200$ MPa; (ii) the material is ductile and $\sigma_{yp} = 250$ MPa.

Solution steps:

(1) For pure shear of $\tau = 100$ MPa, $\sigma_{P1} = -\sigma_{P2} = 100$ MPa, $\sigma_{P3} = 0$
(2) For (i), using the maximum normal stress criterion, FoS = 200/100 = 2
(3) For (ii), using Tresca criterion,

$$\text{FoS} = \frac{\sigma_{yp/2}}{\tau_{\text{max abs.}}} = \frac{250/2}{100} = 1.25$$

Using von Mises criterion,

$$\sigma_{VM} = \sqrt{\sigma_{P_1}^2 + \sigma_{P_2}^2 - \sigma_{P_1}\sigma_{P_2}} = \sqrt{3\sigma_{P_1}^2} = 100\sqrt{3}\text{MPa}$$

$$\text{FoS} = \frac{\sigma_{yp}}{\sigma_{VM}} = \frac{250}{100\sqrt{3}} = 1.44$$

(Note that Tresca criterion is more conservative than von Mises criterion – this is true not just for pure shear but for all states of plane stress).

Problem 13: Find FoS for the pressure vessel of Problem 3 using (i) Tresca criterion (ii) von Mises criterion. Take $\sigma_{yp} = 200$MPa.

Solution steps:

(1) For point A, $\sigma_{P1} = 50$ MPa, $\sigma_{P2} = 35.4$ MPa, $\sigma_{P3} = 0$ leading to
$\tau_{\text{max.abs.}} = \frac{\sigma_{P_1}}{2} = 25\,\text{MPa}$; $\sigma_{VM} = \sqrt{\sigma_{P_1}^2 + \sigma_{P_2}^2 - \sigma_{P_1}\sigma_{P_2}}\,\text{MPa} = 44.53$ MPa
For point C or D,

$$\sigma_{P_1}, \sigma_{P_2} = \frac{50+25}{2} \pm \sqrt{\left(\frac{50-25}{2}\right)^2 + 2.08^2} = 50.17, 24.83\,\text{MPa}$$

along with $\sigma_{P_3} = 0$, leading to

$$\tau_{\text{max.abs.}} = \frac{\sigma_{P_1}}{2} = 25.09\text{MPa}, \quad \sigma_{VM} = 43.45\,\text{MPa}$$

Thus, points *C* and *D* are more severely stressed as compared to point *A* as per Tresca criterion, while the converse is true as per von Mises criterion.
(2) $\text{FoS}_{\text{Tresca}} = \frac{200/2}{25.09} = 3.99$; $\text{FoS}_{\text{VM}} = \frac{200}{44.53} = 4.49$

Problem 14: A circular shaft is subjected to an axial compressive force of 25 kN along with a torque of 300 Nm. Find the minimum diameter required so as to have a FoS of 2 as per Tresca criterion if σ_{yp}=200 MPa.

Solution steps:

(1) If d is the diameter in mm, the state of stress on a critical element located on the outer surface consists of an axial stress and a shear stress given by

$$\sigma = -\frac{25000}{\frac{\pi d^2}{4}} = -\frac{31.83 \times 10^3}{d^2} \text{ MPa}$$

$$\tau = \frac{300 \times 10^3}{\frac{\pi d^3}{16}} = \frac{15.28 \times 10^5}{d^3} \text{ MPa}$$

(2) $\sigma_{P_1}, \sigma_{P_2} = \frac{\sigma}{2} \pm \sqrt{\left(\frac{\sigma}{2}\right)^2 + \tau^2}, \sigma_{P_3} = 0$

Here σ_{P1} is positive and σ_{P2} is negative; hence $\tau_{maxabs} = \left|\frac{\sigma_{P_1} - \sigma_{P_2}}{2}\right|$ and this should be $\leq \frac{\sigma_{yp}/2}{\text{FoS}}$.

Thus, $\left(\frac{\sigma}{2}\right)^2 + \tau^2 \leq 50^2$

i.e. $\frac{15.92^2 \times 10^6}{d^4} + \frac{15.28^2 \times 10^{10}}{d^6} \leq 50^2$

which can be solved by trial and error to yield $d_{min} = 31.82$ mm.

Long Columns 9

9.1 Column

A column is a straight member.

- subjected to uniform or non-uniform compression along its longitudinal axis, and
- likely to fail by buckling (excessive deformation in a mode other than axial compression) at loads well below the axial crushing load.

The discussion here is confined to the case wherein the column bends and such bending is not accompanied or precluded by twisting, and the stresses are within the elastic limit. Such elastic, global flexural buckling occurs for long columns with high torsional rigidity and with the cross-sectional shape either doubly symmetric, or mono-symmetric with the minor principal axis of inertia (corresponding to I_{minimum}) normal to the line of symmetry.[1] Further, any initial imperfection (crookedness of the centroidal axis) or load eccentricity considered here is such that it causes bending about the minor axis when the column is compressed.

9.2 Perfect and Imperfect Columns

An *ideal* or *perfect column* is one with a perfectly straight centroidal axis and with the resultant load applied exactly along that axis. This is never achieved in practice, but is often considered for theoretical analysis because the resulting buckling load is the same

[1] Columns with other cross-sectional shapes undergo pure torsional buckling or flexural–torsional buckling. If the column is thin-walled and not very long, local and distortional buckling modes are also possible.

K. Bhaskar and T. K. Varadan, *Strength of Materials*,
https://doi.org/10.1007/978-3-031-06377-0_9

as the limiting load that would cause excessive bending deflections of the corresponding practical, imperfect column (with an initially crooked centroidal axis due to unavoidable material inhomogeneity and manufacturing errors).

When a perfect column is subjected to compressive loading with some eccentricity with respect to the centroidal axis, the behaviour is similar to that of the imperfect column in that both are problems of stable equilibrium wherein bending occurs along with axial compression right from the onset of loading and the bending deflections corresponding to any value of the compressive load may be uniquely solved for. In both cases, the bending deflections are nonlinearly related to the axial compressive load and grow more and more rapidly as the load is increased and tend to infinity as the load approaches a critical value; this trend is also seen in a *beam-column* subjected to transverse loads together with axial compression. In order to capture this *critical load* in classical perfect column analysis with loading along the centroidal axis, one has to distinguish between states of stable, neutral, and unstable equilibrium by visualizing a small transverse perturbation as described below.

When the compressed perfect column is slightly perturbed (with the axial loading unchanged in magnitude or direction), it is easy to imagine that the bent column would return to the original straight position when the compressive load is small and would diverge away when it is large (see Fig. 9.1 which shows a simply supported column).

Thus, there is a demarcating critical value of the load P_{cr} at which there would be no tendency for the bent column to either return or diverge away. Correspondingly, the compressed and straight prebuckled state is said to be one of *stable equilibrium* when $P < P_{cr}$, *unstable equilibrium* when $P > P_{cr}$, and *neutral equilibrium* when $P = P_{cr}$. These states are shown on the load deflection plot in Fig. 9.1 where the vertical axis corresponds to the straight configuration and the points on the horizontal line at $P = P_{cr}$

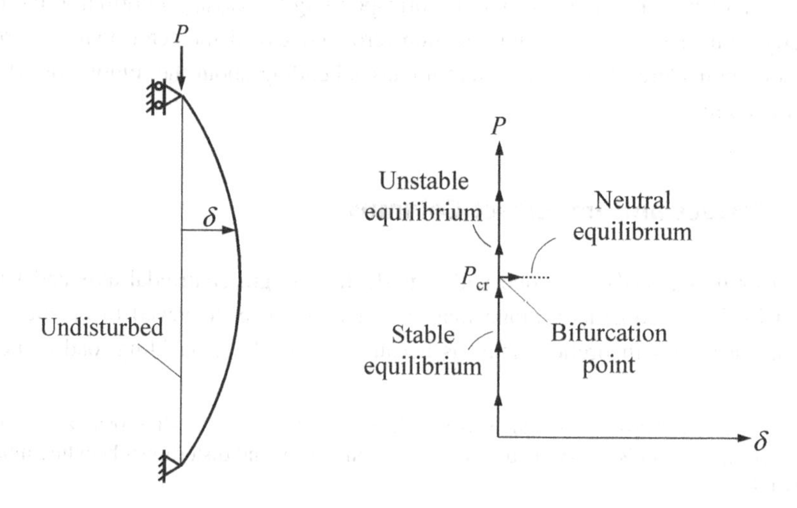

Fig. 9.1 Buckling of a perfect simply supported column

correspond to perturbed bent states; the point where these two paths of equilibrium meet is called the bifurcation point.

The critical load P_{cr} is identified in classical perfect column analysis as the load at which an adjacent, bent equilibrium configuration is possible.

9.3 Assumptions

- All the assumptions of engineering beam theory as stated in Sect. 6.1.
- The critical stress σ_{cr}, i.e. the critical load per unit cross-sectional area is smaller than the elastic limit of the material. This corresponds to elastic buckling which occurs for long, slender columns; short columns undergo inelastic buckling and are not considered here.
- Similarly, in cases of initial imperfection or eccentric loading, the net stresses due to compression and bending are within the elastic limit.

9.4 Final Formulae

$$P_{cr} = \frac{\pi^2 EI}{L_{eq}^2}; \quad \sigma_{cr} = \frac{P_{cr}}{A} = \frac{\pi^2 E}{\left(L_{eq}/r\right)^2}$$

for a prismatic column subjected to end compression, where I is the minimum moment of inertia of the cross-section, r is the corresponding radius of gyration given by $\sqrt{I/A}$, and L_{eq} is the *equivalent length* defined as the length of the corresponding simply supported column having the same buckling load.

$\left(L_{eq}/r\right)$ is called the *slenderness ratio* of the column.

$$\begin{aligned} L_{eq} &= L \text{ for S–S column.} \\ &= L/2 \text{ for C–C column.} \\ &= 2L \text{ for C-F column.} \\ &= 0.699L \approx L/\sqrt{2} \text{ for C-S column.} \end{aligned}$$

where L is the actual physical length and S, C, and F denote simply supported, clamped, and free ends, respectively.

Note that P_{cr} is dependent on the Young's modulus of the material besides the geometric proportions, and not on strength parameters (σ_{yp} or σ_{ult}).

9.5 Illustrative Problems

Note that the formulation is with reference to the deformed configuration in all these problems.

Problem 1: Obtain P_{cr} for an ideal S–S column under end loading.

Solution steps:

(1) Consider the slightly perturbed bent configuration as shown, with deflections highly exaggerated for the sake of clarity.

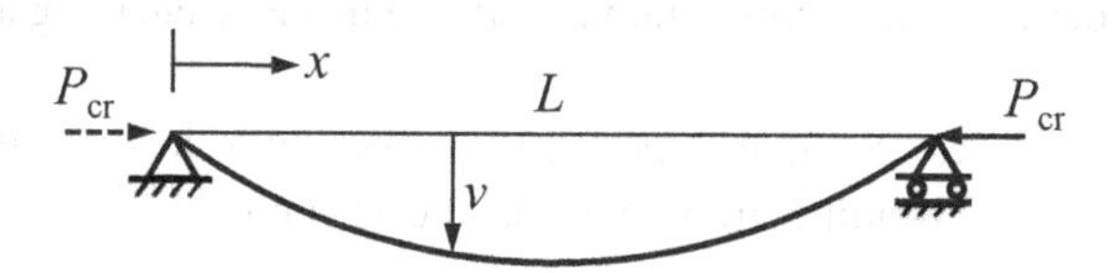

(2) Since the perturbed state is one of equilibrium when $P = P_{cr}$, internal and external moments are equal.

Thus,

$$EIv_{,xx} = -M(x) = -P_{cr}$$

$$v_{,xx} + k^2 v = 0 \text{ where } k = \sqrt{\frac{P_{cr}}{EI}}$$

$$v = A\cos kx + B\sin kx$$

(3) Using zero deflection conditions at the ends,

$$v(0) = 0 \Rightarrow A = 0$$

$$v(L) = 0 \Rightarrow B\sin kL = 0 \Rightarrow B = 0 \text{ OR } \sin kL = 0$$

If both A and B are zero, v itself would be zero and this simply corresponds to the prebuckled straight position being a state of equilibrium.
Discarding this *trivial solution*, one has

$$\sin kL = 0$$

which is referred to as the *characteristic buckling equation*. This is a transcendental equation with infinite roots as given by

$$kL = n\pi$$

$$\Rightarrow P_{cr} = \frac{n^2\pi^2 EI}{L^2}, \quad n = 1, 2, \ldots$$

with $\sin kx = \sin \frac{n\pi x}{L}$, $n = 1, 2, \ldots$ as the corresponding buckled shapes.

Thus, this is an eigenvalue problem with a non-trivial solution only for specific, discrete values of P_{cr}; these values are referred to as eigenvalues and the associated buckled shapes as mode shapes.

(4) The lowest root of the characteristic equation alone is of practical relevance because the corresponding load is the limiting load for an actual imperfect column as stated earlier. Thus, confining attention to the lowest root alone and denoting the corresponding load alone as P_{cr} (here as well as in later problems involving perfect columns), one has, for the S–S column,

$$P_{cr} = \frac{\pi^2 EI}{L^2}$$

with $\sin \frac{\pi x}{L}$ as the mode shape. This load is often referred to as the *Euler load.*

Note that the amplitude B is undetermined—it can be any small value so as to correspond to an adjacent state of neutral equilibrium as shown in Fig. 9.1.

Problem 2: Obtain P_{cr} by considering an eccentrically loaded ideal S-S column as shown. Also find the maximum compressive stress due to any load P. For what values of P can one analyse it as a beam subjected to end-moments and calculate the deflections and bending stresses without incurring much error?

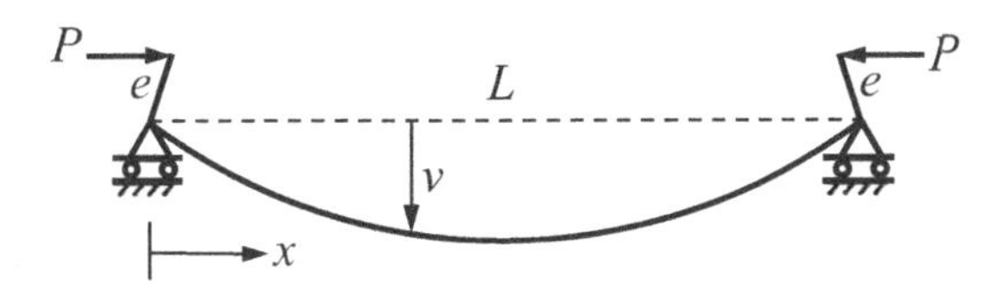

Solution steps:

(1) For any value of P, the bending moment $M(x) = P(e + v)$
Thus, $EIv_{,xx} = -P(e + v)$

$$v_{,xx} + \lambda^2 v = -\lambda^2 e \quad \text{where } \lambda = \sqrt{\frac{P}{EI}} \quad \left(\text{i. e. } \lambda L = \pi\sqrt{\frac{P}{P_{cr}}}\right)$$

$$v = A\cos\lambda x + B\sin\lambda x - e$$

(2) $v(0) = v(L) = 0 \Rightarrow A = e,\ B = \frac{e(1-\cos\lambda L)}{\sin\lambda L}$

i.e. $v = e(\cos\lambda x - 1) + \frac{e(1-\cos\lambda L)\sin\lambda x}{\sin\lambda L}$

(3) $v_{\max} = v\left(\frac{L}{2}\right) = e\left(\cos\frac{\lambda L}{2} - 1\right) + \frac{e(1-\cos\lambda L)\sin\frac{\lambda L}{2}}{\sin\lambda L}$

This can be simplified as

$$v_{\max} = e\left(\sec\frac{\lambda L}{2} - 1\right)$$
$$= k_1\left(\frac{PeL^2}{8EI}\right), \quad \text{with } k_1 = \left(\frac{\sec\frac{\lambda L}{2} - 1}{(\lambda L)^2/8}\right)$$

where $\left(\frac{PeL^2}{8EI}\right)$ is the central deflection of a S–S beam due to end-moments Pe alone.

Thus, k_1 represents the second-order beam-column effect (due to the term Pv in the bending moment) on the deflection—it monotonically increases with P and tends to infinity as $\lambda L \to \pi$, or as $P \to P_{cr}\left(= \frac{\pi^2 EI}{L^2}\right)$; the actual variation is as shown below.

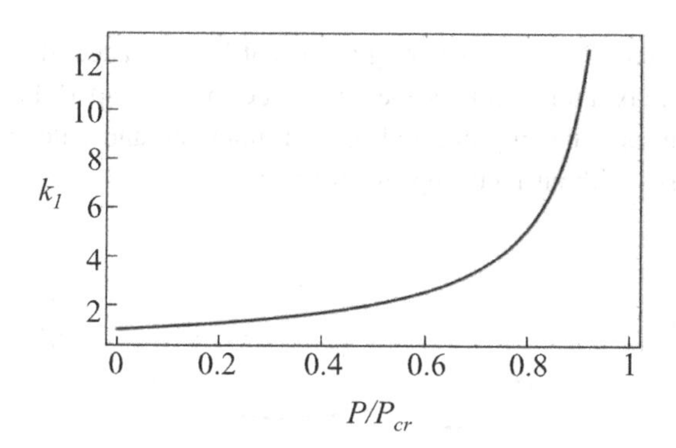

(4) For any value of P, the maximum stress occurs at mid-span and at the extreme fibre on the compression side. If c is the corresponding distance from the neutral axis,

$$|\sigma_{\max}| = \frac{P}{A} + \frac{P(e + v_{\max})c}{I} = \frac{P}{A} + \frac{Pec\sec\frac{\lambda L}{2}}{I}$$
$$= \frac{P}{A}\left(1 + \frac{ec}{r^2}\sec\left(\frac{L}{2r}\sqrt{\frac{P}{EA}}\right)\right)$$

where r is the radius of gyration corresponding to I.

(This equation, called the *secant formula*, is used to find out the allowable P corresponding to which $\sigma_{\max}$ reaches the elastic limit for a given column and a specified load eccentricity e. Because of the transcendental nature of the equation, a graphical approach is often used for this purpose.)

(5) Noting that $\frac{Pec}{I}$ is the bending stress due to the end-moments Pe alone, the factor $\left(\sec\frac{\lambda L}{2}\right)$ represents the corresponding second-order beam-column effect. For values of

P/P_{cr} less than 1/30, both this factor and k_1 are very close (within 5%) to unity. Thus, for such small values of P as compared to the critical load, the second-order beam-column effect may be neglected while calculating deflections and bending stresses (as was done earlier in Problem 1 of Chap. 8.)

Problem 3: Obtain P_{cr} by considering the beam-column problem shown. For what values of the axial force P can one neglect its effect on deflections and bending stresses?

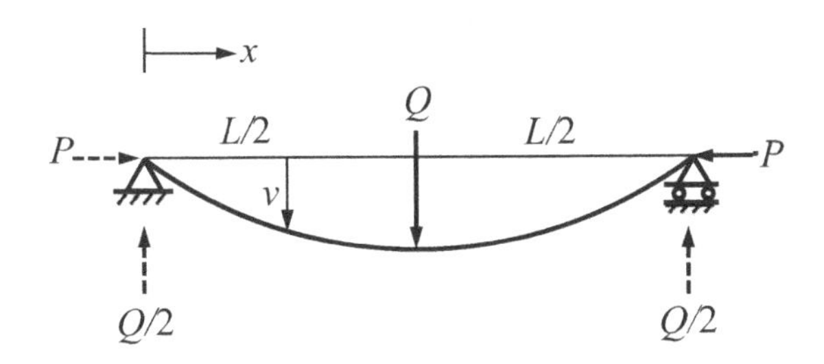

Solution steps:

(1) $M(x) = Pv + \frac{Qx}{2}$ in the interval (0, L/2)

(2) $EIv_{,xx} = -Pv - \frac{Qx}{2}$

i.e. $v_{,xx} + \lambda^2 v = -\frac{Qx}{2EI}$ where $\lambda = \sqrt{\frac{P}{EI}}$ (i. e. $\lambda L = \pi\sqrt{\frac{P}{P_{cr}}}$)

$v = A\cos\lambda x + B\sin\lambda x - \frac{Qx}{2P}$

(3) $v(0) = v_{,x}\left(\frac{L}{2}\right) = 0 \Rightarrow A = 0,\ B = \dfrac{Q}{2P\lambda\cos\frac{\lambda L}{2}}$

$$v_{\max} = v\left(\frac{L}{2}\right) = \frac{Q}{2P\lambda}\left(\tan\frac{\lambda L}{2} - \frac{\lambda L}{2}\right)$$

$$= k_2\left(\frac{QL^3}{48EI}\right), \quad \text{with } k_2 = \left(\frac{3\left(\tan\frac{\lambda L}{2} - \frac{\lambda L}{2}\right)}{\left(\frac{\lambda L}{2}\right)^3}\right)$$

where $\frac{QL^3}{48EI}$ is the value corresponding to transverse load alone, and the factor k_2 represents the effect of the compressive force. The variation of this factor with P is very similar (see Problem 6) to that of k_1 in Problem 2, leading to infinitely large deflections as $P \rightarrow P_{cr}\left(= \frac{\pi^2 EI}{L^2}\right)$.

(4) The maximum bending stress corresponds to the bending moment occurring at mid-span as given by

$$M_{\max} = \frac{QL}{4} + Pv_{\max} = \frac{QL}{4}\left(1 + \frac{k_2 PL^2}{12EI}\right) = \frac{QL}{4}\left[1 + \frac{k_2(\lambda L)^2}{12}\right]$$

where $\frac{QL}{4}$ is the value corresponding to transverse load alone, and the factor in brackets the influence of the axial force P. This factor, and its counterpart k_2 for the maximum deflection, are very close (within 4%) to unity when P/P_{cr} is less than 1/30.

Thus, for such small values of P, deflections and bending stresses may be obtained by considering the transverse load alone; the axial stress P/A is then superposed to get the net normal stress distribution across the cross-section.

Problem 4: Obtain P_{cr} by considering an imperfect S-S column as shown.

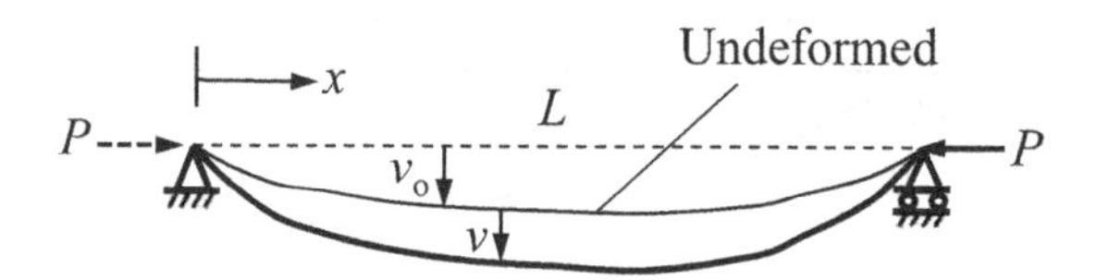

Solution steps:

(1) The initial undeformed shape may be expressed in Fourier series as

$$v_o = \sum_m A_m \sin \frac{m\pi x}{L}$$

(2) Noting that the additional curvature at any point is related to the bending moment and that this is obtained in terms of the deflection $v(x)$ measured from the undeformed position, one has

$$EIv_{,xx} = -P(v_o + v)$$

i.e. $v_{,xx} + \lambda^2 v = -\lambda^2 v_o$ where $\lambda = \sqrt{\frac{P}{EI}}$ (i. e. $\lambda L = \pi\sqrt{\frac{P}{P_{cr}}}$)

(3) The complete solution satisfying $v(0) = v(L) = 0$ is given by

$$v = \sum_m \frac{\lambda^2 A_m}{\left(\frac{m\pi}{L}\right)^2 - \lambda^2} \sin \frac{m\pi x}{L}$$

which shows that the deflections tend to infinity as $\lambda \rightarrow \pi/L$, i.e. as $P \rightarrow P_{cr}\left(= \frac{\pi^2 EI}{L^2}\right)$.

Problem 5: For the practical, initially crooked column of Problem 4, show that a nearly linear relationship exists between δ and δ/P where δ is the central deflection corresponding to P, measured from the undeformed position.

Solution steps:

(1) Noting that the initial shape, of nature as shown in the figure for Prob.4, is likely to be the one commonly encountered in practice, it is to be expected that the corresponding Fourier coefficients A_m (see Step 1 therein) would decrease as m increases.
(2) For $P < P_{cr}$, λ is less than π/L, and hence the denominator in the series for v (Step 3 of Prob.4) increases as m increases.
(3) In view of the above observations, the first term in the series for v is very dominant as compared to the other terms – especially so if P is not too small compared to P_{cr}.
(4) Thus, by neglecting higher terms, one can write

$$v \approx \frac{\lambda^2 A_1}{\left(\frac{\pi}{L}\right)^2 - \lambda^2} \sin \frac{\pi x}{L}$$

i.e. $v_{\text{cen}} = \delta \approx \frac{\lambda^2 A_1}{\left(\frac{\pi}{L}\right)^2 - \lambda^2} = \frac{A_1}{\left(\frac{\pi}{\lambda L}\right)^2 - 1} = \frac{A_1}{\frac{P_{cr}}{P} - 1}$.

i.e. $\frac{\delta}{P} P_{cr} - \delta \approx A_1$.

which is the equation of a straight line as shown with its inverse slope equal to P_{cr} and with the x-intercept equal to A_1, a measure of the initial crookedness of the column.

(This is referred to as *Southwell plot* and it provides a convenient and safe non-destructive method to directly obtain the critical load of a practical column with unknown material properties; it also yields a quantitative estimate of the initial imperfection without any actual measurements. In such a test, the column is loaded gradually and well below its critical load such that the transverse deflections from the undeformed position are small. While obtaining the best straight line fit, points corresponding to higher magnitudes of P are taken as more important.)

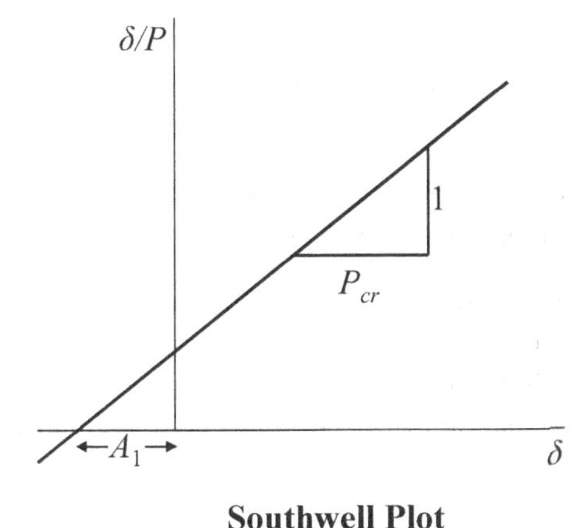

Southwell Plot

Problem 6: Show that the deflection magnification factors due to beam-column effect — k_1 for the eccentrically loaded column in Problem 2 and k_2 for the beam-column in Problem 3 — are well approximated by the factor $\frac{1}{\left(1-\frac{P}{P_{cr}}\right)}$. Further, discuss the importance of this factor with respect to the imperfect S–S column of Problems 4 and 5.

Solution steps:

(1) A tabular comparison of the factors k_1, k_2 and the suggested approximate factor is presented below.

$P\,/\,P_{cr}$	0.1	0.2	0.4	0.6	0.8
k_1	1.11	1.26	1.69	2.55	5.12
k_2	1.11	1.25	1.66	2.48	4.94
$\frac{1}{(1-P/P_{cr})}$	1.11	1.25	1.67	2.50	5.00

Thus, the suggested factor is a good approximation over the entire allowable range of loading.

(2) With reference to the imperfect column subjected to load P not too small as compared to P_{cr}, a good estimate of the central deflection v_{cen} measured from the undeformed, initially crooked position is given by $\left(\frac{A_1}{\frac{P_{cr}}{P}-1}\right)$ as discussed in Problem 5.
If the initial shape is close to a half-sine wave (a likely possibility in practice), A_1 itself is a good estimate of the initial central deviation from the straight line joining the centroids of the end cross-sections (see the figure given for Problem 4) since the higher coefficients A_2, etc. would be very small in comparison.
The final central deviation from the straight line is thus

$$v_{\text{cen}} + A_1 = \frac{A_1}{\frac{P_{cr}}{P} - 1} + A_1 = \frac{A_1}{\left(1 - \frac{P}{P_{cr}}\right)}$$

Thus, the initial central deviation of the column gets amplified by the approximate factor $\frac{1}{\left(1-\frac{P}{P_{Cr}}\right)}$ after loading. The maximum bending moment is calculated by using the final deviation as the moment arm for the axial force P, and thus the suggested approximate factor is useful for determining $P_{\text{allowable}}$ corresponding to any specified maximum allowable stress (after A_1 itself is found out by the Southwell Plot experiment or by careful measurement of the initial imperfection).

Problem 7: Obtain P_{cr} and the corresponding mode shape for an ideal (a) C–F column; (b) C–S column; (c) C–C column; and (d) S–S column with rotation at the ends restrained by torsional springs of stiffness $k_t = 4EI/L$.

Solution steps for (a):

(1) Let δ be the deflection at the free end of the buckled column as shown.

(2) $EIv_{,xx} = -M = P_{cr}(\delta - v)$

$$v_{,xx} + k^2 v = k^2 \delta \quad \text{where } k = \sqrt{\frac{P_{cr}}{EI}}$$

$$v = A\cos kx + B\sin kx + \delta$$

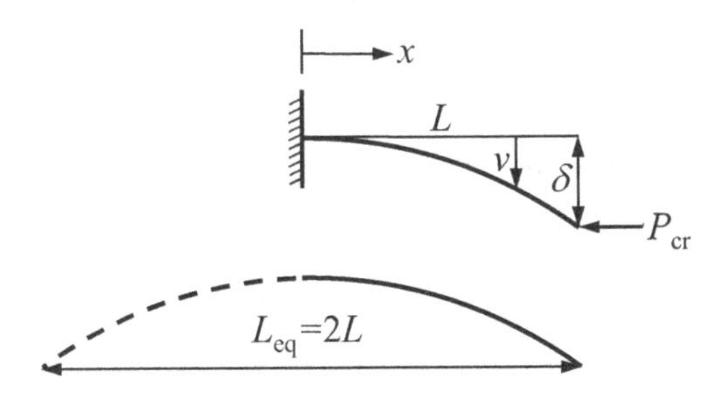

(3) Using $v(0) = v_{,x}(0) = 0$, $v(L) = \delta$, one gets, for a non-trivial solution,

$$\cos kL = 0 \Rightarrow kL = \frac{\pi}{2} \Rightarrow P_{cr} = \frac{\pi^2 EI}{4L^2} = \frac{\pi^2 EI}{(2L)^2}$$

with $\left(1 - \cos\frac{\pi x}{2L}\right)$ as the mode shape. This shape is essentially one-half of that of a simply supported column of length $2L$ as shown, and thus $L_{eq} = 2\,L$ for this case.

Solution steps for (b):

(1) In the buckled configuration as shown, the fixed end moment M_o can be balanced only by the set of transverse forces R for equilibrium.

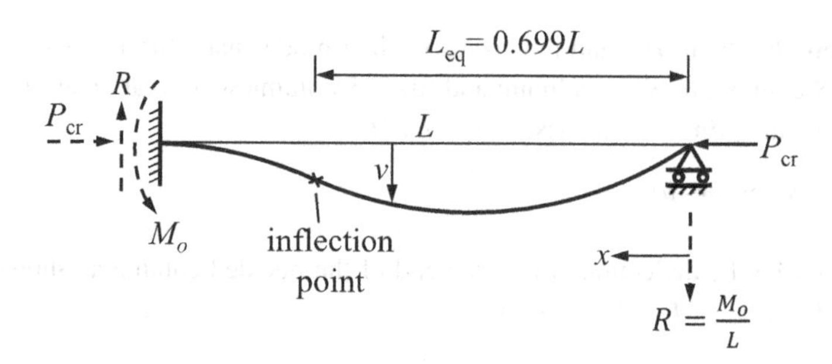

Thus $M = P_{cr}v - Rx$, with x measured from the hinge as shown.

(2) Proceeding as before, $v = A\cos kx + B\sin kx + \frac{Rx}{k^2EI}$

(3) Using $v(0) = v(L) = v_{,x}(L) = 0$, one gets, for a non-trivial solution,

$$\tan kL - kL = 0$$

This transcendental equation may be solved by trial and error or graphical methods to yield

$$kL = 4.493 \Rightarrow P_{cr} = \frac{20.19EI}{L^2} = \frac{\pi^2 EI}{(0.699L)^2}$$

with the mode shape given by

$$\sin\left(4.493\frac{x}{L}\right) - \left(\frac{x}{L}\right)\sin 4.493$$

Solution steps for (*c*):

(1) Theoretically, the column may buckle into a shape symmetric about mid-span or an antisymmetric shape with zero deflection and curvature at mid-span; in the latter case, the problem reduces to that of a C–S column of length *L*/2, and thus P_{cr} would be $\frac{20.19EI}{(L/2)^2} = \frac{80.76EI}{L^2}$.
In the following steps, the symmetric shape is considered.

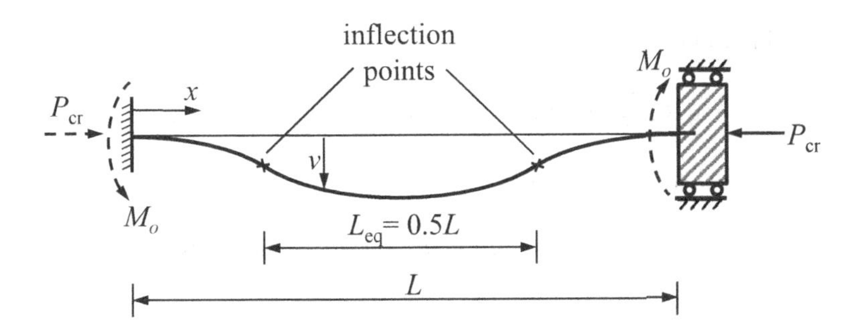

(2) With reference to the figure, $M = P_{cr}v - M_o$.
(3) Proceeding as before, $v = A\cos kx + B\sin kx + \frac{M_o}{k^2EI}$
(4) Using $v(0) = v_{,x}(0) = v_{,x}(L/2) = 0$, one getscritical load corresponding to the symmetric mode shape is

$$\sin\frac{kL}{2} = 0 \Rightarrow kL = 2\pi$$

Thus, $P_{cr} = \frac{4\pi^2EI}{L^2} = \frac{\pi^2EI}{(L/2)^2}$ with$\left(1 - \cos\frac{2\pi x}{L}\right)$ as the mode shape.
(5) The critical load corresponding to the symmetric mode shape is lower than that of the antisymmetric shape, and is hence the required final answer.
(6) As shown, this mode shape includes a simple half-sine-wave portion between the two inflection points at which curvature is zero; thus this central portion corresponds to a S–S column of length $L/2$ and this is the equivalent length L_{eq} for the C–C column.

Solution steps for (d):

(1) Since both S–S and C–C columns have the lowest buckling load corresponding to a symmetric shape about mid-span, the same is expected here.

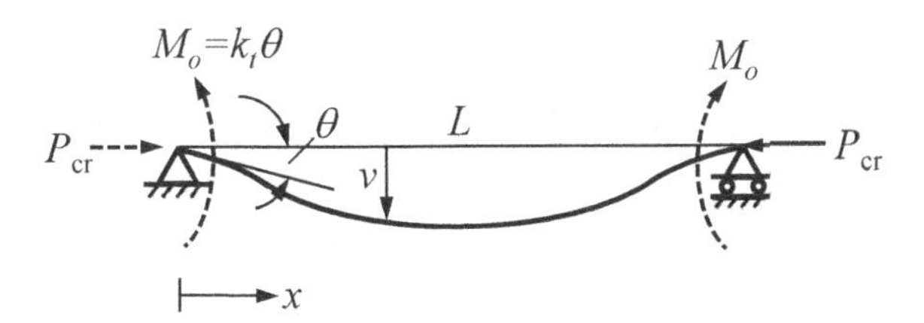

(2) The solution steps are the same as for C–C column except that $v_{,x}(0)$ is nonzero here and given by

$$v_{,}(0) = \frac{M_o}{k_t} = \frac{M_oL}{4EI}$$

This yields

$$4\sin\frac{kL}{2} + kL\cos\frac{kL}{2} = 0 \Rightarrow kL = 4.578 \Rightarrow P_{cr} = \frac{20.96EI}{L^2}$$

with the mode shape given by

$$1 - \cos\left(4.578\frac{x}{L}\right) + 1.145\sin\left(4.578\frac{x}{L}\right)$$

Problem 8: For a long column of thin-walled equal angle section as shown, find P_{cr} assuming the ends to be simply supported.

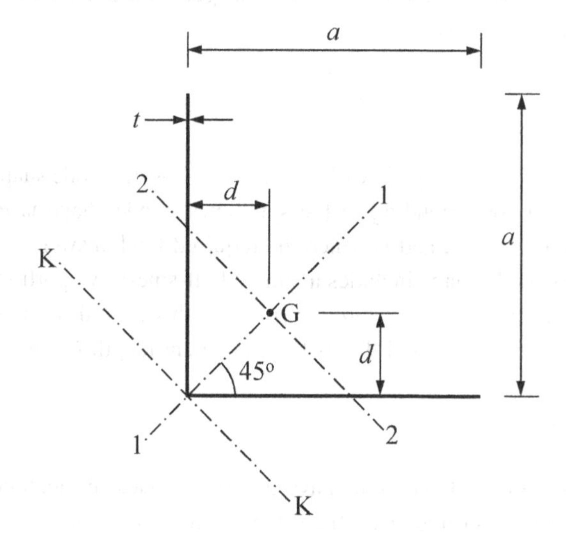

Solution steps:

(1) The distance d of the centroid G from the left or bottom leg is calculated as

$$d = \frac{ta \cdot \frac{a}{2}}{2ta} = \frac{a}{4}$$

(2) The axis of symmetry 1–1 and the axis 2–2 normal to it and passing through G are the principal axes of the cross-section. Visual inspection clearly reveals that 2–2 is the minor axis corresponding to $I_{\min}$ and the column would undergo pure flexural buckling by bending about it.

Considering an axis K-K as shown and noting that the two legs of the angle section contribute equally to I_{KK}, one gets

$$I_{KK} = 2\int_0^a t\left(s\cos 45°\right)^2 ds = \frac{ta^3}{3}$$

with s coordinate starting from the intersection point.

Hence, I_{min} about axis $2-2 = I_{KK} - 2ta(d\sqrt{2})^2 = \frac{ta^3}{12}$

(One should note that calculation of I_{max} about axis 1–1 involves the same steps as that of I_{KK}, and thus $I_{max} = 4I_{min}$.)

(3) Thus, $P_{cr} = \frac{\pi^2 EI_{min}}{L^2}$ for S–S column

$= \frac{\pi^2 Eta^3}{12L^2}$

Problem 9: For the stepped column shown, find P_{cr}.

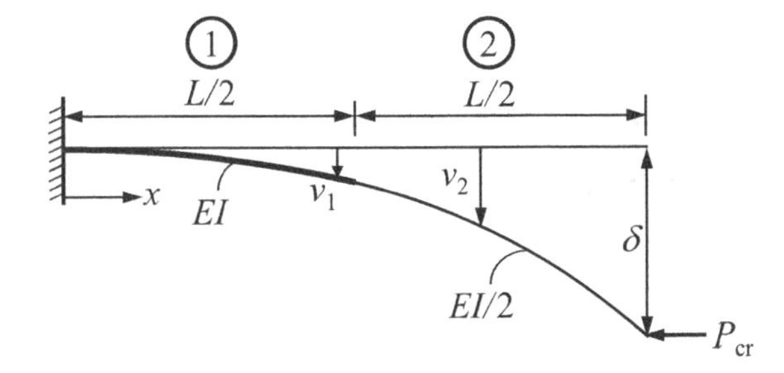

Solution steps:

(1) The two regions should be analysed separately with appropriate continuity conditions at the junction.

(2) Proceeding as done for the prismatic cantilever of Problem 7(a),

$$v_{1,xx} + k^2 v_1 = k^2\delta \quad \text{where } k = \sqrt{\frac{P_{cr}}{EI}}$$

$$\Rightarrow v_1 = A_1 \cos kx + B_1 \sin kx + \delta$$

$$v_{2,xx} + 2k^2 v_2 = 2k^2\delta$$

$$\Rightarrow v_2 = A_2 \cos\sqrt{2}kx + B_2 \sin\sqrt{2}kx + \delta$$

(3) Using $v_1(0) = v_{1,x}(0) = 0$, $v_2(L) = \delta$ along with

$$v_1\left(\frac{L}{2}\right) = v_2\left(\frac{L}{2}\right), \quad v_{1,x}\left(\frac{L}{2}\right) = v_{2,x}\left(\frac{L}{2}\right),$$

one gets, after due simplification,

$$\sin\frac{kL}{2} - \sqrt{2}\cos\frac{kL}{2}\cot\frac{kL}{\sqrt{2}} = 0 \Rightarrow kL = 1.438$$

$$\Rightarrow P_{cr} = \frac{2.067EI}{L^2}$$

Problem 10: For the S-S column subjected to two equal axial loads as shown, find P_{cr}.

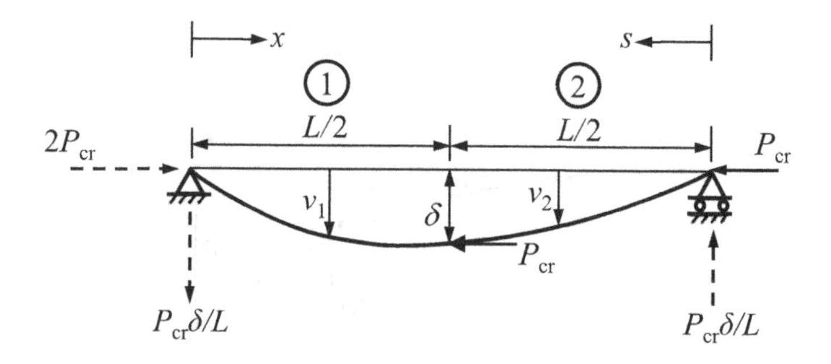

Solution steps:

(1) If δ is the central deflection, then it is clear that transverse reactions of $P_{cr}\delta/L$ at the supports are required for satisfying moment equilibrium.
(2) In terms of x-coordinate for the left half and s coordinate for the right half as shown,

$$EIv_{1,xx} = -M_1(x) = -2P_{cr}v_1 + \frac{P_{cr}\delta x}{L}$$

$$v_{1,xx} + 2k^2v_1 = \frac{k^2\delta x}{L} \text{ where } k = \sqrt{\frac{P_{cr}}{EI}}$$

$$v_1 = A_1\cos\sqrt{2}kx + B_1\sin\sqrt{2}kx + \frac{\delta x}{2L}$$

$$EIv_{2,ss} = -M_2(s) = -P_{cr}v_2 - \frac{P_{cr}\delta s}{L}$$

$$v_{2,ss} + k^2v_2 = -\frac{k^2\delta s}{L}$$

$$v_2 = A_2 \cos ks + B_2 \sin ks - \frac{\delta s}{L}$$

(3) Using

$$v_1(0) = v_2(0) = 0,\ v_1\left(\frac{L}{2}\right) = v_2\left(\frac{L}{2}\right) = \delta$$

along with
$v_{1,x}\left(\frac{L}{2}\right) = -v_{2,s}\left(\frac{L}{2}\right)$ (minus sign since x and s are in opposite directions),

one gets

$$3kL\left(\cot\frac{kL}{2} + \frac{\cot\frac{kL}{\sqrt{2}}}{\sqrt{2}}\right) - 1 = 0$$

$$\Rightarrow kL = 2.557 \Rightarrow P_{cr} = \frac{6.536EI}{L^2}$$

Problem 11: For a column made of steel, with $E = 200$ GPa and elastic limit = 200 MPa, find the slenderness ratio below which inelastic buckling would occur. What is the corresponding $\left(\frac{L_{eq}}{d}\right)$ ratio if the cross-section is
(a) a circle of diameter d;
(b) rectangle of size b x d with $b > d$?

Solution steps:

(1) $\sigma_{cr} = \frac{\pi^2 E}{(L_{eq}/r)^2} \Rightarrow \frac{L_{eq}}{r} = \pi\sqrt{\frac{E}{\sigma_{cr}}}$

(2) When σ_{cr} reaches the elastic limit, $\frac{L_{eq}}{r} = \pi\sqrt{\frac{200\times10^9}{200\times10^6}} \approx 100$. Columns shorter than this would undergo inelastic buckling.

(3) For (a), $r = \sqrt{\frac{I}{A}} = \sqrt{\frac{\pi d^4/64}{\pi d^2/4}} = \frac{d}{4}$

For (b), $r = \sqrt{\frac{I_{\min}}{A}} = \sqrt{\frac{bd^3/12}{bd}} = \frac{d}{3.46}$.

Thus, the limiting $\frac{L_{eq}}{d}$ ratios are approximately 25 and 29, respectively.

Energy Methods 10

The application of Castigliano's theorems for displacement analysis is illustrated here with reference to structures involving bar, circular shaft, and beam elements of the types discussed in earlier chapters.

10.1 Assumptions

- Linearly elastic material behaviour.
- A linear relationship between any generalized displacement (linear or angular) and any generalized load (force or moment or torque), so that the strain energy of the structure is a quadratic function of the generalized loads.

10.2 Final Formulae

$$U_{\text{bar}} = \int \frac{P^2 \mathrm{d}x}{2AE}; \quad U_{\text{shaft}} = \int \frac{T^2 \mathrm{d}x}{2GJ}; \quad U_{\text{beam}} = \int \frac{M^2 \mathrm{d}x}{2EI}$$

$$U = \sum \frac{F_i^2 L_i}{2A_i E_i} \text{ for a pin jointed truss with member forces } F_i$$

$$\frac{\partial U}{\partial Q_i} = q_i \text{(Castigliano's Theorem II)}$$

$$\frac{\partial U}{\partial R_i} = 0 \text{ (Castigliano's theorem of least work for indeterminate structures)}$$

K. Bhaskar and T. K. Varadan, *Strength of Materials*,
https://doi.org/10.1007/978-3-031-06377-0_10

where U is the strain energy, Q_i is a generalized load and q_i is the corresponding generalized displacement in the same direction, and R_i is a redundant support reaction/moment or an internal redundant force/moment.

Note that the strain energy of the beam does not include any contribution from the shear force V, in accordance with the classical assumption of neglect of the corresponding shear strain.

For members subjected to combined loading, with attention confined here to those with circular or axisymmetric annular cross-section, the total strain energy is obtained by identifying the variations of the axial force P along the centroidal axis, the net bending moment M, and the torque T, and by simply adding the corresponding U_{bar}, U_{beam}, and U_{shaft}.

10.3 Illustrative Problems

The following points should be noted.

- Each applied load is given a separate algebraic name to facilitate differentiation, and its value is substituted only at the end.
- When a generalized displacement other than those corresponding to the applied loads is desired, an appropriate dummy load is introduced.
- While evaluating the derivative of U with respect to a load, it is often convenient to do so before summation or integration.
- In statically indeterminate problems, after the redundant support reactions and/or internal forces are determined, they are looked upon as known applied loads on the corresponding determinate structure for further displacement analysis (see Problems 10 and 11).
- In problems of combined loading involving long members, the strain energy due to axial loading may be neglected (see Problem 14).

Problem 1: Find the axial displacements of the sections at C, D, and F.

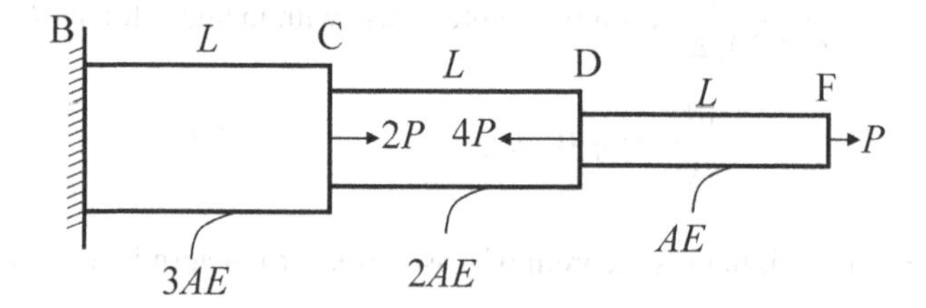

Solution steps:

(1) Denote the loads by Q_1 (=P), Q_2 (=$-$ $4P$), Q_3 (=$2P$) as shown.

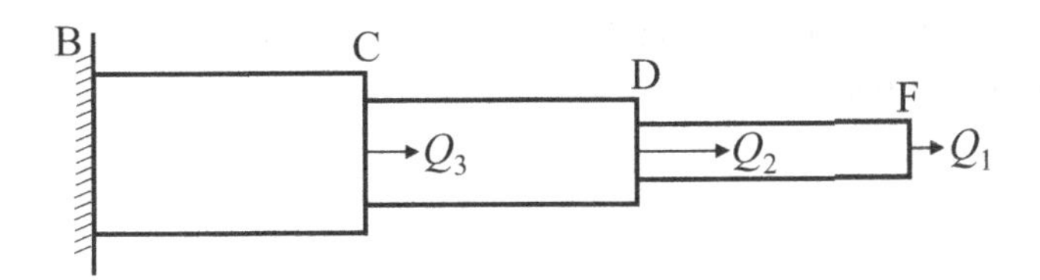

(2) The tensile force acting on DF is Q_1, that on CD is $(Q_1 + Q_2)$ and that on BC is $(Q_1 + Q_2 + Q_3)$.

(3) $U = \dfrac{Q_1^2 L}{2AE} + \dfrac{(Q_1 + Q_2)^2 L}{2(2AE)} + \dfrac{(Q_1 + Q_2 + Q_3)^2 L}{2(3AE)}$

(4) Disp. at $F = q_1 = \dfrac{\partial U}{\partial Q_1}$

$$= \frac{Q_1 L}{AE} + \frac{(Q_1 + Q_2)L}{2AE} + \frac{(Q_1 + Q_2 + Q_3)L}{3AE} = \frac{\left(P - \frac{3P}{2} - \frac{P}{3}\right)L}{AE} = -\frac{5PL}{6AE}$$

$$\text{Disp.at } D = q_2 = \frac{\partial U}{\partial Q_2}$$
$$= \frac{(Q_1 + Q_2)L}{2AE} + \frac{(Q_1 + Q_2 + Q_3)L}{3AE}$$
$$= \frac{\left(-\frac{3P}{2} - \frac{P}{3}\right)L}{AE} = -\frac{11PL}{6AE}$$

$$\text{Disp. at } C = q_3 = \frac{\partial U}{\partial Q_3} = \frac{(Q_1 + Q_2 + Q_3)L}{3AE} = -\frac{PL}{3AE}$$

where the minus signs indicate that these displacements are leftwards.

Problem 2: Same as Problem 1 except that the area of cross-section of BC varies linearly from $2A$ at C to $4A$ at B, and displacement at F alone is sought.

Solution steps:

(1) In terms of a local x-coordinate from C to B, $A_{BC} = 2A\left(1 + \dfrac{x}{L}\right)$

(2) $U = \dfrac{Q_1^2 L}{2AE} + \dfrac{(Q_1 + Q_2)^2 L}{2(2AE)} + \displaystyle\int_0^L \dfrac{(Q_1 + Q_2 + Q_3)^2 \mathrm{d}x}{2E \, \mathrm{x} \, 2A\left(1 + \frac{x}{L}\right)}$

$$= \frac{Q_1^2 L}{2AE} + \frac{(Q_1 + Q_2)^2 L}{2(2AE)} + \frac{(Q_1 + Q_2 + Q_3)^2 L \ln 2}{2(2AE)}$$

$$(3)\ q_1 = \frac{\partial U}{\partial Q_1} = \frac{Q_1 L}{AE} + \frac{(Q_1 + Q_2)L}{2AE} + \frac{(Q_1 + Q_2 + Q_3)L \ln 2}{2AE}$$
$$= \frac{\left(P - \frac{3P}{2} - \frac{P \ln 2}{2}\right)L}{AE} = -\frac{(1 + \ln 2)PL}{2AE}$$

Problem 3: If CG and HI are rigid links and links BC and GJ have the same *AE*, find the displacement of point I.

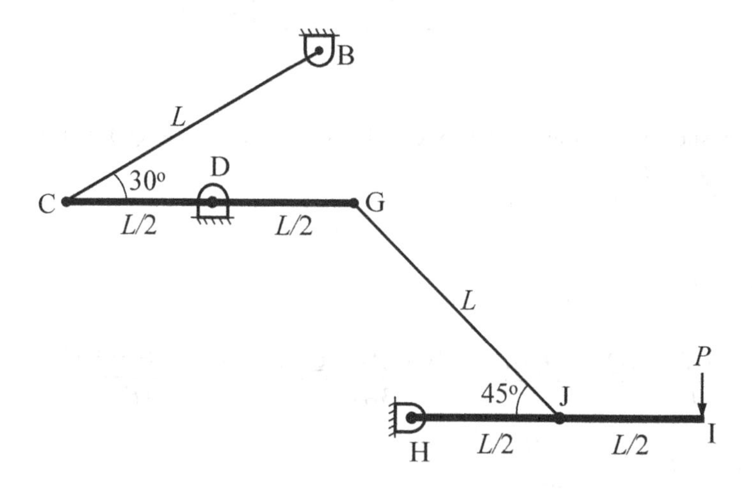

Solution steps:

(1) As the horizontal rigid link HI rotates about H, point I moves vertically downwards, and this is the generalized displacement corresponding to load P. Thus, it is necessary to find $\frac{\partial U}{\partial P}$.

(2) The rigid links do not store strain energy and so their member forces are not required. Due to applied loading, F_{GJ} is tensile and F_{BC} is compressive.

(3) Considering forces on HI,

$$\text{moments about H} = 0 \Rightarrow F_{\mathrm{GJ}} \frac{L}{2} \sin 45^\circ = PL \Rightarrow F_{\mathrm{GJ}} = 2\sqrt{2}P$$

(4) Similarly, by taking moment of forces on CG about D, one gets

$$F_{\mathrm{GJ}} \frac{L}{2} \sin 45^\circ = |F_{\mathrm{BC}}| \frac{L}{2} \sin 30^\circ \Rightarrow F_{\mathrm{BC}} = -4P$$

$$(5)\ U = \frac{F_{\mathrm{BC}}^2 L}{2AE} + \frac{F_{\mathrm{GJ}}^2 L}{2AE} = \frac{24P^2 L}{2AE} \Rightarrow \delta_{\mathrm{I}} = \frac{\partial U}{\partial P} = \frac{24PL}{AE} \downarrow$$

Problem 4: Find the torque reactions at the two supports.

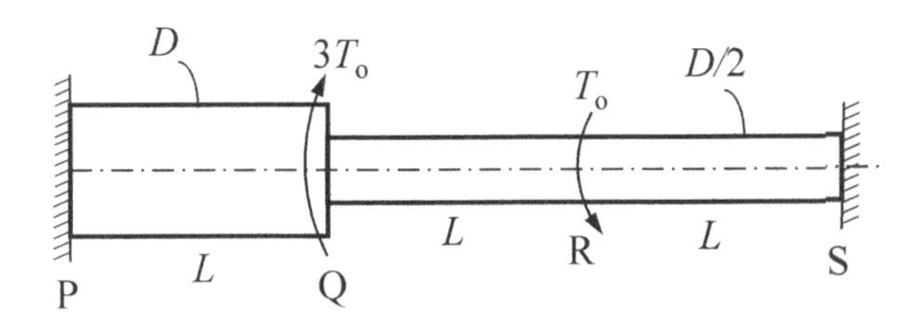

Solution steps:

(1) This is an externally statically indeterminate problem with just one equilibrium equation and two unknown support reactions. Denote the applied torques as $T_Q = 3T_o$ and $T_R = -T_o$, and take the reaction T_S at S as the redundant, all in the same sense for convenience.

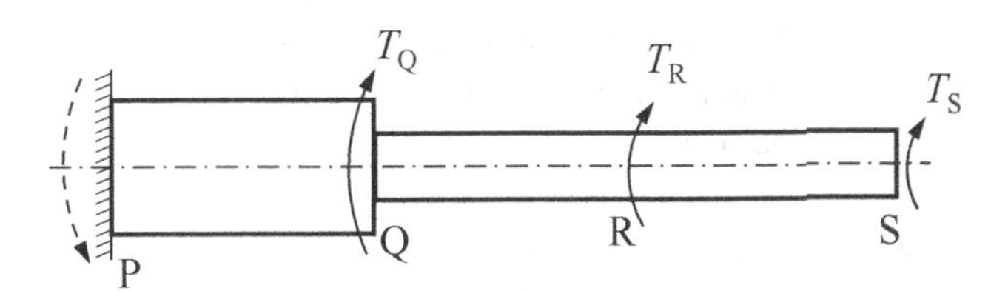

Then, segment RS is subjected to torque T_S, QR to $(T_S + T_R)$, and PQ to $(T_S + T_R + T_Q)$.

(2) $U = \dfrac{T_S^2 L}{2GJ} + \dfrac{(T_S + T_R)^2 L}{2GJ} + \dfrac{(T_S + T_R + T_Q)^2 L}{2GJ_1}$

where $J = \dfrac{\pi\left(\frac{D}{2}\right)^4}{32}$, $J_1 = \dfrac{\pi D^4}{32}$.

(3) $\dfrac{\partial U}{\partial T_S} = 0 \Rightarrow \dfrac{T_S L}{GJ} + \dfrac{(T_S + T_R)L}{GJ} + \dfrac{(T_S + T_R + T_Q)L}{GJ_1} = 0 \Rightarrow T_S = \dfrac{14T_o}{33}$

(4) Reaction at $P = |T_S + T_R + T_Q| = \frac{80T_o}{33}$ in the opposite sense as shown.

Problem 5: Find the support reactions.

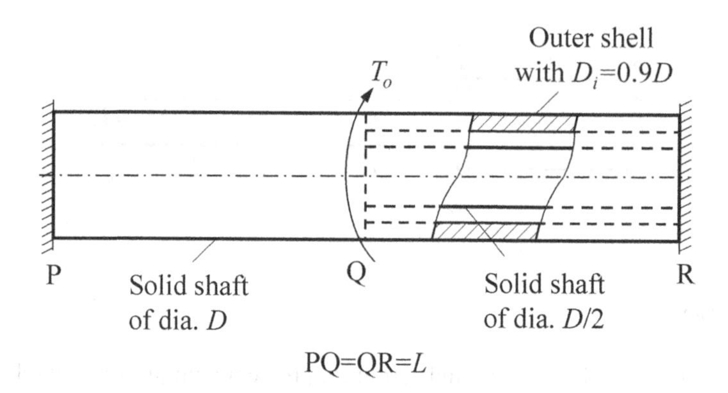

Solution steps:

(1) Besides being externally statically indeterminate, this is also internally indeterminate because the distribution of the torque in the segment QR between the central shaft and the outer shell is unknown. Let T_R be the reaction at R and T_S be the portion of it exerted on the outer shell, both in the same sense as T_0; these are the chosen redundants.
Thus, over the length QR, the central shaft is subjected to torque $(T_R - T_S)$ and the outer shell to T_S, while segment PQ is subjected to $(T_R + T_o)$.

(2) $U = \frac{T_S^2 L}{2GJ_S} + \frac{(T_R - T_S)^2 L}{2GJ_C} + \frac{(T_R + T_o)^2 L}{2GJ_{PQ}}$
where $J_S = \frac{\pi D^4(1-0.9^4)}{32}$, $J_C = \frac{\pi\left(\frac{D}{2}\right)^4}{32}$, $J_{PQ} = \frac{\pi D^4}{32}$.

(3) $\frac{\partial U}{\partial T_S} = 0 \Rightarrow \frac{T_S}{J_S} = \frac{T_R - T_S}{J_C} \Rightarrow T_S = 0.846 T_R$

(4) $\frac{\partial U}{\partial T_R} = 0 \Rightarrow \frac{T_R - T_S}{J_C} = -\frac{(T_R + T_o)}{J_{PQ}} \Rightarrow T_R = -0.289 T_o$
where the minus sign indicates that it is opposite to the sense of T_o.

(5) Hence, $T_P = 0.711 T_o$ in the sense opposite to T_o.

Problem 6: Find the central deflection. (This was solved by moment-area method in Chap. 7, Problem 9).

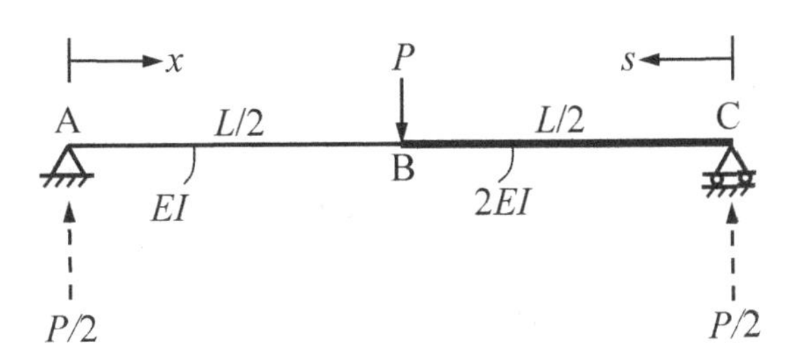

Solution steps:

(1) In terms of the coordinates x and s as shown,

$$M = \frac{Px}{2} \text{ in AB}, \quad \frac{Ps}{2} \text{ in CB}$$

(2) $U = \int_0^{L/2} \frac{\left(\frac{Px}{2}\right)^2}{2EI} \mathrm{d}x + \int_0^{L/2} \frac{\left(\frac{Ps}{2}\right)^2}{2(2EI)} \mathrm{d}s = \frac{P^2L^3}{128EI}$

(3) $v_{\text{cen}} = \frac{\partial U}{\partial P} = \frac{PL^3}{64EI} \downarrow$

Problem 7: Find the prop reaction.

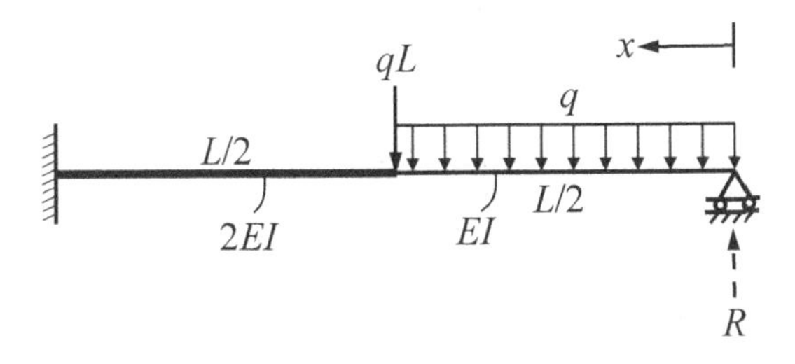

Solution steps:

(1) In terms of the prop reaction R as shown,

$$M(x) = \left(Rx - \frac{qx^2}{2}\right) \text{in}\left(0, \frac{L}{2}\right)$$
$$\text{and} \left[Rx - \frac{qL}{2}\left(x - \frac{L}{4}\right) - qL\left(x - \frac{L}{2}\right)\right] \text{in}\left(\frac{L}{2}, L\right)$$

(2) $U = \int_0^{L/2}\left(\frac{M^2}{2EI}\right)\mathrm{d}x + \int_{L/2}^{L}\left(\frac{M^2}{2(2EI)}\right)\mathrm{d}x$

(3) $$\frac{\partial U}{\partial R} = \int_0^{L/2}\left(\frac{MM_{,R}}{EI}\right)\mathrm{d}x + \int_{L/2}^{L}\left(\frac{MM_{,R}}{2EI}\right)\mathrm{d}x$$

$$= \int_0^{L/2}\left(\frac{Mx}{EI}\right)\mathrm{d}x + \int_{L/2}^{L}\left(\frac{Mx}{2EI}\right)\mathrm{d}x = 0$$

$$\Rightarrow \left(\frac{RL^3}{24EI} - \frac{qL^4}{128EI}\right) + \left(\frac{7RL^3}{48EI} - \frac{19qL^4}{384EI} - \frac{5qL^4}{96EI}\right) = 0 \Rightarrow R = \frac{7qL}{12}\uparrow$$

Problem 8: In Problem 3, find the angle of rotation of CG about D.

Solution steps:

(1) The formal procedure involves three steps:
 (i) At D, apply a dummy moment M_o on link CG as shown.
 (ii) Find member forces due to this along with the actual load P, and calculate U.
 (iii) Find the rotation of CG as $\left.\frac{\partial U}{\partial M_o}\right|_{M_o=0}$.

$$\text{Thus,}\ \ \theta_{\mathrm{CG}} = \left.\left\{\frac{\partial}{\partial M_o}\left(\frac{F_{\mathrm{BC}}^2 L}{2AE} + \frac{F_{\mathrm{GJ}}^2 L}{2AE}\right)\right\}\right|_{M_o=0}$$
$$= \left(\frac{F_{\mathrm{BC},M_o}L}{AE}\right)F_{\mathrm{BC}}|_{M_o=0} + \left(\frac{F_{\mathrm{GJ},M_o}L}{AE}\right)F_{\mathrm{GJ}}|_{M_o=0}$$

which shows that the dummy moment is required only to find the member force derivatives F_{BC,M_o} and F_{GJ,M_o}, and that this can always be done separately if desired.

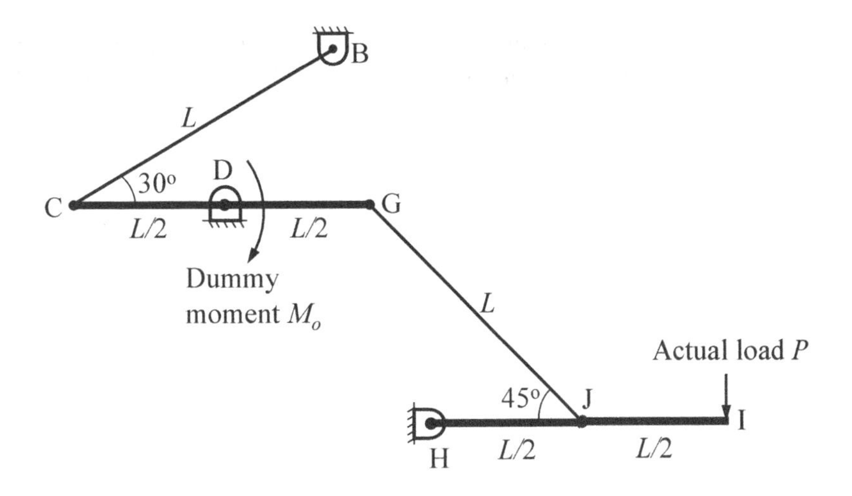

In other words, as an alternative to the formal procedure, one can find all the member forces due to the actual loading alone as a first step. Then, one can apply the dummy load alone, find the member force derivatives, and use them to find the generalized displacement corresponding to the dummy load.

(2) Due to M_o alone, it is clear that only BC (and the portion CD of the rigid bar) will resist it, while the remaining members remain unstressed.

Thus, $|F_{\mathrm{BC}}|\dfrac{L}{2}\sin 30^\circ = M_o \Rightarrow F_{\mathrm{BC}} = -\dfrac{4M_o}{L} \Rightarrow F_{\mathrm{BC},M_o} = -\dfrac{4}{L}$

while $F_{\mathrm{GJ},M_o} = 0$.

(From the above, it is clear that the member force derivatives with respect to a dummy load may be calculated as the member forces themselves due to unit magnitude of that dummy load,

i.e. $F_{\mathrm{BC},M_o} = F_{\mathrm{BC}}$ *due to unit* M_o, *and so on.*)

(3) So, $\theta_{\mathrm{CG}} = \left(\frac{F_{\mathrm{BC},M_o}L}{AE}\right) F_{\mathrm{BC}}|_{M_o=0}$

$= \left(-\frac{4}{AE}\right)(-4P$ from Prob.3) $= \frac{16P}{AE}$ rad. in the clockwise direction (same as that of the dummy moment M_o).

(Note that this problem can also be solved by considering a vertical dummy load at C or G and by dividing the corresponding linear displacement by $L/2$.)

Problem 9: Assuming same *AE* for all members, find (i) the vertical displacement at B and the horizontal displacement at *C*; (ii) the horizontal displacement at *B*.

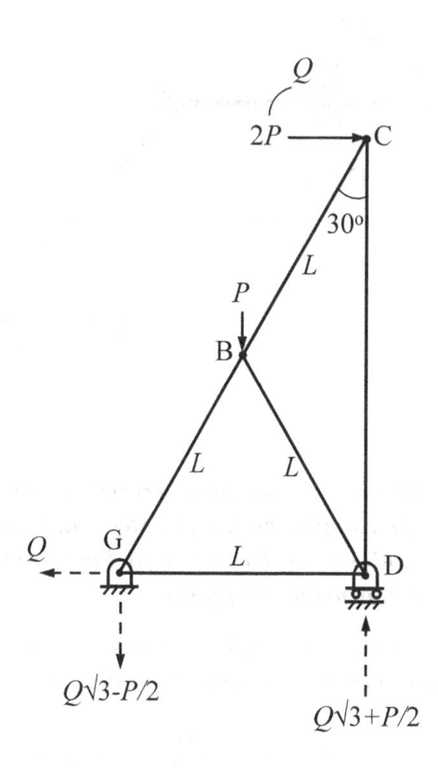

Solution steps for (i):

(1) The load at *C* is called *Q*.

(2) Considering overall equilibrium, left and right support reactions are calculated as shown.

(3) Considering equilibrium of each pin joint, member forces F_i are found out.

(4) $U = \frac{\sum F_i^2 L_i}{2AE}$

(5) δ_v at $B = \frac{\partial U}{\partial P} = \frac{\sum F_i F_{i,P} L_i}{AE}$

δ_h at $C = \frac{\partial U}{\partial Q} = \frac{\sum F_i F_{i,Q} L_i}{AE}$

All calculations are conveniently carried out in tabular form as shown below.

Mem	L_i	F_i	$F_{i,P}$	$F_{i,Q}$	$F_iF_{i,P}L_i$	$F_iF_{i,Q}L_i$
GB	L	$2Q-\frac{P}{\sqrt{3}}$	$-\frac{1}{\sqrt{3}}$	2	$\left(\frac{P}{3}-\frac{2Q}{\sqrt{3}}\right)L$	$\left(4Q-\frac{2P}{\sqrt{3}}\right)L$
GD	L	$\frac{P}{2\sqrt{3}}$	$\frac{1}{2\sqrt{3}}$	0	$\frac{PL}{12}$	0
BD	L	$-\frac{P}{\sqrt{3}}$	$-\frac{1}{\sqrt{3}}$	0	$\frac{PL}{3}$	0
BC	L	$2Q$	0	2	0	$4QL$
CD	$L\sqrt{3}$	$-Q\sqrt{3}$	0	$-\sqrt{3}$	0	$3\sqrt{3}QL$
				Sum	$\left(\frac{3P}{4}-\frac{2Q}{\sqrt{3}}\right)L$	$\left(8+3\sqrt{3}\right)QL$ $-\frac{2PL}{\sqrt{3}}$

Hence, δ_v at $B=\left(\frac{3P}{4}-\frac{2Q}{\sqrt{3}}\right)\frac{L}{AE}=\left(\frac{3}{4}-\frac{4}{\sqrt{3}}\right)\frac{PL}{AE}\downarrow$

$$\delta_h \text{ at } C=\left(8+3\sqrt{3}\right)\frac{QL}{AE}-\frac{2PL}{\sqrt{3}AE}=\frac{16\left(3+\sqrt{3}\right)}{3}\frac{PL}{AE} \text{ to the right.}$$

Solution steps for (*ii*):

(1) Find the member forces due to actual loads P and Q as above.
(2) Introduce a dummy unit horizontal force R at B as shown.
(3) Due to this alone, find the member forces to get

$$F_{\mathrm{GB},R}=-F_{\mathrm{BD},R}=1,\quad F_{\mathrm{GD},R}=\frac{1}{2},\quad F_{\mathrm{BC},R}=F_{\mathrm{CD},R}=0$$

4. δ_h at $B=\frac{\partial U}{\partial R}\Big|_{R=0}=\sum\left(\frac{F_{i,R}\,L_i}{\mathrm{AE}}\,F_i|_{R=0}\right)$
$=\left[\left(2Q-\frac{P}{\sqrt{3}}\right)+\frac{1}{2}\frac{P}{2\sqrt{3}}-\left(-\frac{P}{\sqrt{3}}\right)\right]\frac{L}{AE}=\left(\frac{16\sqrt{3}+1}{4\sqrt{3}}\right)\frac{PL}{AE}$ to the right.

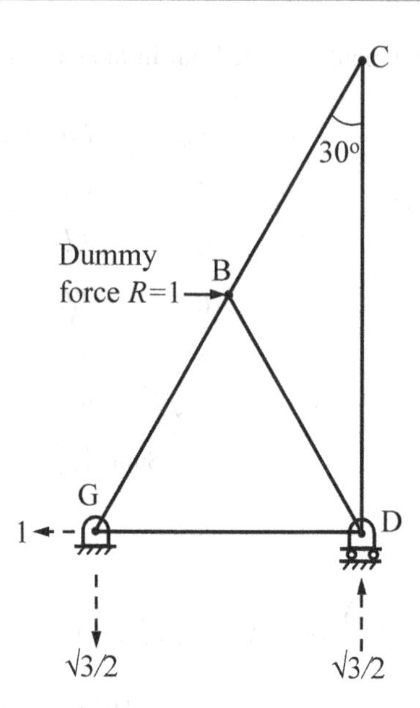

Problem 10: Assuming same AE for all members, find (i) the horizontal and vertical displacements at B; (ii) horizontal displacement at G.

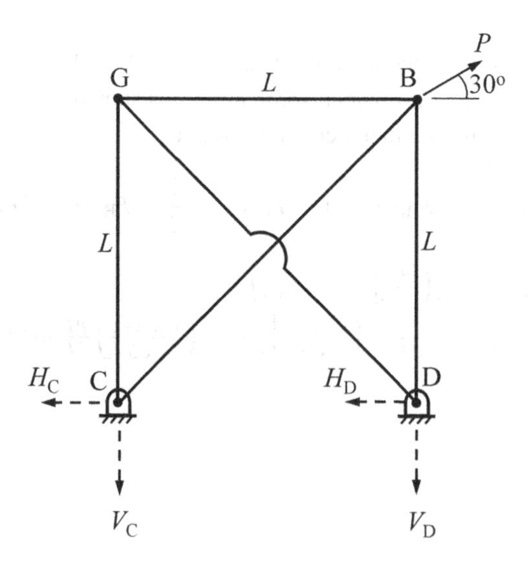

Solution steps:

(1) In order to get the two displacement components at B, replace the inclined load P by a horizontal force P_h (=P cos 30°) and a vertical force P_v (=P sin 30°).
(2) For four support reactions H_C, V_C, H_D, and V_D, there are only three equations of statics—hence, externally statically indeterminate.
(3) Taking H_D as a redundant, one gets

$$V_C = P_h, \quad V_D = P_v - P_h, \quad H_C = P_h - H_D$$

(4) Considering equilibrium of each pin joint, member forces are obtained as

$$F_{BD} = P_v - P_h + H_D; \quad F_{GD} = -H_D\sqrt{2};$$

$$F_{GB} = F_{GC} = H_D; \quad F_{BC} = \sqrt{2}(P_h - H_D)$$

(5) $U = \left(F_{GB}^2 + F_{GC}^2 + F_{BD}^2\right)\frac{L}{2AE} + \left(F_{GD}^2 + F_{BC}^2\right)\frac{L\sqrt{2}}{2AE}$

$$\frac{\partial U}{\partial H_D} = 0 = \left(F_{GB}F_{GB,H_D} + F_{GC}F_{GC,H_D} + F_{BD}F_{BD,H_D}\right)\frac{L}{AE} + \left(F_{GD}F_{GD,H_D} + F_{BC}F_{BC,H_D}\right)\frac{L\sqrt{2}}{AE}$$

$$\Rightarrow [F_{GB}(1) + F_{GC}(1) + F_{BD}(1)]\frac{L}{AE} + \left[F_{GD}\left(-\sqrt{2}\right) + F_{BC}\left(-\sqrt{2}\right)\right]\frac{L\sqrt{2}}{AE} = 0$$

$$\Rightarrow H_D = \frac{\left(1 + 2\sqrt{2}\right)P_h - P_v}{\left(3 + 4\sqrt{2}\right)} = \frac{\left(1 + 2\sqrt{2}\right)P\cos 30^o - P\sin 30^o}{\left(3 + 4\sqrt{2}\right)} = 0.325P$$

(6) ***For further analysis, the structure is converted into a statically determinate one as shown with the support restraint corresponding to H_D removed and with H_D applied as a new load besides P_h and P_v; this released structure has the same member forces as in Step 4 and the same U as in Step 5. The three loads are now considered independent generalized loads, and their values in terms of P will be substituted only after obtaining the final expressions for the required displacements.***

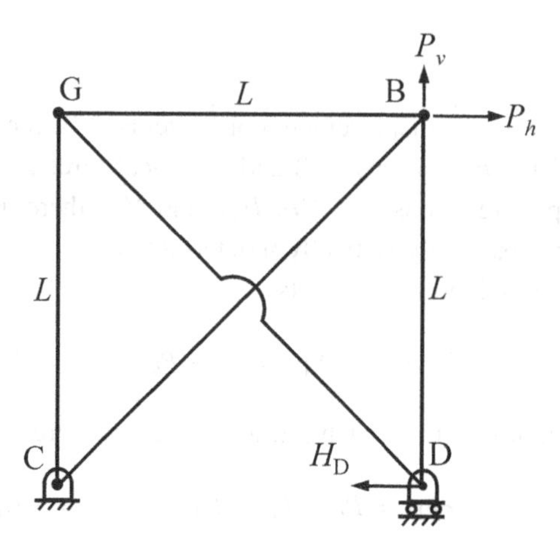

Released structure

(7) Thus,

$$\delta_{hB} = \frac{\partial U}{\partial P_h} = \left(F_{GB}F_{GB,P_h} + F_{GC}F_{GC,P_h} + F_{BD}F_{BD,P_h}\right)\frac{L}{AE} + \left(F_{GD}F_{GD,P_h} + F_{BC}F_{BC,P_h}\right)\frac{L\sqrt{2}}{AE}$$

$$\Rightarrow \delta_{hB} = [F_{GB}(0) + F_{GC}(0) + F_{BD}(-1)]\frac{L}{AE} + \left[F_{GD}(0) + F_{BC}\left(\sqrt{2}\right)\right]\frac{L\sqrt{2}}{AE}$$

$$\Rightarrow \delta_{hB} = \left[\left(1 + 2\sqrt{2}\right)(P_h - H_D) - P_v\right]\frac{L}{AE}$$

After substituting the values of P_h, P_v, and H_D, one gets

$$\delta_{hB} = 1.57\frac{PL}{AE} \text{ to the right.}$$

Similarly, $\delta_{vB} = \frac{\partial U}{\partial P_v} = (P_v - P_h + H_D)\frac{L}{AE} = -0.0408\frac{PL}{AE}$, with the minus sign indicating that it is downwards.

It can easily be verified that these final results are the same as those that would be obtained by first substituting for H_D in terms of P_h and P_v in the expression for U and then finding the derivatives with respect to P_h and P_v.

(8) To get the horizontal displacement at G, consider the member forces of the released statically determinate structure due to a unit dummy force H_G alone as shown, yielding

$$F_{\mathrm{GB},H_\mathrm{G}} = F_{\mathrm{BD},H_\mathrm{G}} = -1, \quad F_{\mathrm{GD},H_\mathrm{G}} = F_{\mathrm{GC},H_\mathrm{G}} = 0, \quad F_{\mathrm{BC},H_\mathrm{G}} = \sqrt{2}$$

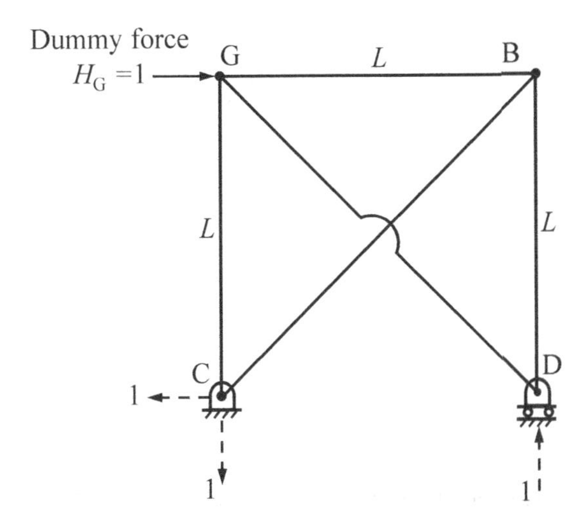

Hence,

$$\begin{aligned}
\delta_{h\mathrm{G}} &= \left.\frac{\partial U}{\partial H_\mathrm{G}}\right|_{H_\mathrm{G}=0} = \sum\left(\frac{F_{i,H_\mathrm{G}} L_i}{AE} F_i|_{H_\mathrm{G}=0}\right) \\
&= (-F_\mathrm{GB} - F_\mathrm{BD})\frac{L}{AE} + \left(F_\mathrm{BC}\sqrt{2}\right)\frac{L\sqrt{2}}{AE} \\
&= \left[(1+2\sqrt{2})P_h - P_v - 2(1+\sqrt{2})H_\mathrm{D}\right]\frac{L}{AE} \\
&= 1.245\frac{PL}{AE} \text{ to the right}
\end{aligned}$$

Problem 11: Same as Problem 10 except that D is a roller support and C and D are connected by a truss member with the same AE as other members.

Solution steps:

(1) This is externally statically determinate with the three support reactions found out using the three equations of statics as

$$V_\mathrm{C} = H_\mathrm{C} = P_h, \quad V_\mathrm{D} = P_v - P_h$$

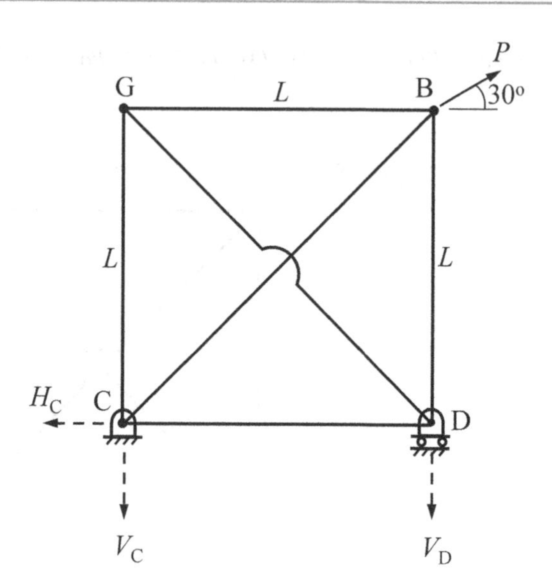

(2) However, it is internally statically indeterminate since three members are connected at each pin joint, and thus, the internal member forces cannot be determined in terms of the applied loads alone. Taking CD as a redundant member and considering equilibrium of each pin joint, other member forces are obtained; their values are the same as in Step 4 of Problem 10 with the tensile force F_{CD} in place of H_D.

(3) $U = \left(F_{GB}^2 + F_{GC}^2 + F_{BD}^2 + F_{CD}^2\right)\frac{L}{2AE} + \left(F_{GD}^2 + F_{BC}^2\right)\frac{L\sqrt{2}}{2AE}$

$$\frac{\partial U}{\partial F_{CD}} = 0$$

$$\Rightarrow \left(F_{GB}F_{GB,F_{CD}} + F_{GC}F_{GC,F_{CD}} + F_{BD}F_{BD,F_{CD}} + F_{CD}\right)\frac{L}{AE}$$

$$+ \left(F_{GD}F_{GD,F_{CD}} + F_{BC}F_{BC,F_{CD}}\right)\frac{L\sqrt{2}}{AE} = 0$$

$$\Rightarrow F_{CD} = \frac{\left(1 + 2\sqrt{2}\right)P_h - P_v}{4\left(1 + \sqrt{2}\right)} = \frac{\left(1 + 2\sqrt{2}\right)P\cos 30^\circ - P\sin 30^\circ}{4\left(1 + \sqrt{2}\right)} = 0.292P$$

with the positive sign indicating that it is tensile.

(4) ***For further analysis, consider the released statically determinate structure as shown, with the redundant member CD removed and with forces F_{CD} applied at C and D, and taken as independent generalized loads besides P_h and P_v.***

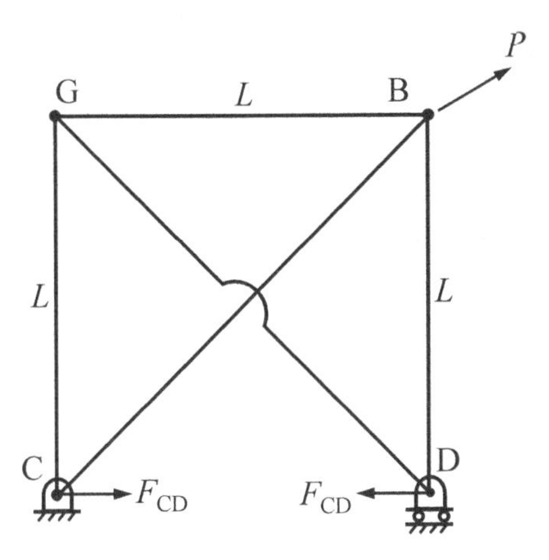

***This released structure has the same member forces as in Step 2 above, but its strain energy is U as obtained in Step 3 minus that corresponding to member* CD.**

(5) Thus,

$$\delta_{h\text{B}} = \frac{\partial U_{\text{released}}}{\partial P_h} = \left(F_{\text{GB}}F_{\text{GB},P_h} + F_{\text{GC}}F_{\text{GC},P_h} + F_{\text{BD}}F_{\text{BD},P_h}\right)\frac{L}{AE} + \left(F_{\text{GD}}F_{\text{GD},P_h} + F_{\text{BC}}F_{\text{BC},P_h}\right)\frac{L\sqrt{2}}{AE}$$

$$\Rightarrow \delta_{h\text{B}} = \left[\left(1 + 2\sqrt{2}\right)(P_h - F_{\text{CD}}) - P_v\right]\frac{L}{AE} = 1.699\frac{PL}{AE} \text{ to the right.}$$

Similarly,

$$\delta_{v\text{B}} = \frac{\partial U_{\text{released}}}{\partial P_v} = (P_v - P_h + F_{\text{CD}})\frac{L}{AE} = -0.0745\frac{PL}{AE}$$

with the minus sign indicating that it is downwards.

It can easily be verified that these final results are the same as those that would be obtained by first substituting for F_{CD} in terms of P_h and P_v in the expression for U of the original statically indeterminate structure and then finding the derivatives with respect to P_h and P_v.

(6) The horizontal displacement at G is obtained exactly as in Step 8 of Problem 10, as

$$\delta_{h\text{G}} = \frac{\partial U_{\text{released}}}{\partial H_{\text{G}}} = (-F_{\text{GB}} - F_{\text{BD}})\frac{L}{AE} + \left(F_{\text{BC}}\sqrt{2}\right)\frac{L\sqrt{2}}{AE}$$
$$= \left[(1 + 2\sqrt{2})P_h - P_v - 2(1 + \sqrt{2})F_{\text{CD}}\right]\frac{L}{AE}$$
$$= 1.408\frac{PL}{AE} \text{ to the right.}$$

Problem 12: For the beam of Prob.6, find the slope at A and the deflection at a distance of $L/4$ from A.

Solution steps:

(1) Introduce a dummy unit moment M_o at the left end leading to support reactions as shown.

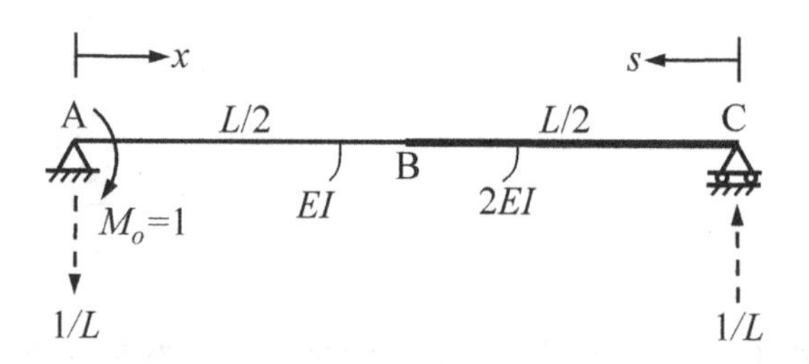

(2) Correspondingly, one gets $M_{,M_o} = \left(1 - \frac{x}{L}\right)$ in AB, and $\frac{s}{L}$ in CB.

(3) $\theta_A = \left.\frac{\partial U}{\partial M_o}\right|_{M_o=0}$

$$= \int_0^{L/2} \left(\frac{M_{,M_o}}{EI}\right) M|_{M_o=0} \mathrm{d}x + \int_0^{L/2} \left(\frac{M_{,M_o}}{2EI}\right) M|_{M_o=0} \mathrm{d}s$$

$$= \int_0^{\frac{L}{2}} \frac{\left(\frac{Px}{2}\right)\left(1 - \frac{x}{L}\right)}{EI} \mathrm{d}x + \int_0^{\frac{L}{2}} \frac{\left(\frac{Ps}{2}\right)\left(\frac{s}{L}\right)}{2EI} \mathrm{d}s$$

$$= \frac{5PL^2}{96EI} \circlearrowright$$

(4) Similarly, with a dummy unit force applied at $x = L/4$ as shown,

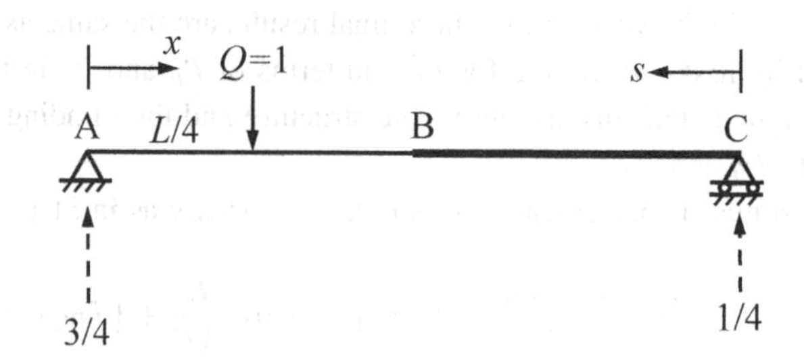

$$
\begin{aligned}
M_{,Q} &= \frac{3x}{4} \text{for } x = 0 \text{ to } \frac{L}{4} \text{ in AB} \\
&= \frac{3x}{4} - 1\left(x - \frac{L}{4}\right), \text{i.e.} \frac{(L-x)}{4} \quad \text{for } x = \frac{L}{4} \text{ to } \frac{L}{2} \text{ in AB} \\
&= \frac{s}{4} \text{in CB.}
\end{aligned}
$$

$$
\begin{aligned}
v_{x=L/4} &= \left.\frac{\partial U}{\partial Q}\right|_{Q=0} \\
&= \int_0^{L/4} \left(\frac{M_{,Q}}{EI}\right) M|_{Q=0} \mathrm{d}x + \int_{L/4}^{L/2} \left(\frac{M_{,Q}}{EI}\right) M|_{Q=0} \, \mathrm{d}x \\
&\quad + \int_0^{L/2} \left(\frac{M_{,Q}}{2EI}\right) M|_{Q=0} \mathrm{d}s \\
&= \frac{3PL^3}{256EI} \downarrow
\end{aligned}
$$

Problem 13: For the beam of Problem 7, find v_{cen} and the slope at the prop.

Solution steps:

(1) Consider the statically determinate structure without the prop and with R applied along with the actual loading as shown. The value of R, as determined in Problem 7, will be substituted at the end.
Further, in order to determine v_{cen}, the corresponding load qL is designated as P.

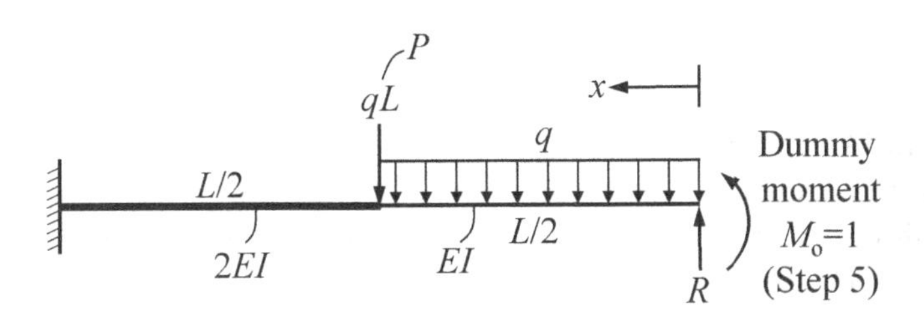

(2) $M(x) = \left(Rx - \frac{qx^2}{2}\right)$ in $\left(0, \frac{L}{2}\right)$

$$\text{and} \left[Rx - \frac{qL}{2}\left(x - \frac{L}{4}\right) - P\left(x - \frac{L}{2}\right)\right] \text{ in } \left(\frac{L}{2}, L\right)$$

(3) $M_{,P} = 0$ in $\left(0, \frac{L}{2}\right)$ and $-\left(x - \frac{L}{2}\right)$ in$\left(\frac{L}{2}, L\right)$

(4) $v_{\text{cen}} = \frac{\partial U}{\partial P} = \int_0^{L/2} 0\,\text{d}x + \int_{L/2}^{L} \left(\frac{MM_{,P}}{2EI}\right)\text{d}x$

$$= \int_{\frac{L}{2}}^{L} -\frac{\left[Rx - \frac{qL}{2}\left(x - \frac{L}{4}\right) - P\left(x - \frac{L}{2}\right)\right]\left(x - \frac{L}{2}\right)}{2EI}\text{d}x$$

$$= \frac{(8P + 7qL - 20R)L^3}{384EI} = \frac{5qL^4}{576EI} \downarrow \text{ (after substitution for } P \text{ and } R)$$

(5) With a dummy unit moment M_o alone applied at the tip of the statically determinate cantilever as shown, one gets

$M(x) = M_o$, and hence, $M_{,M_o} = 1$.

(6) Thus, $\theta_{\text{tip}} = \left.\frac{\partial U}{\partial M_o}\right|_{M_o=0}$

$$= \int_0^{L/2} \left(\frac{1}{EI}\right) M|_{M_o=0}\text{d}x + \int_{L/2}^{L} \left(\frac{1}{2EI}\right) M|_{M_o=0}\text{d}x$$

$$= \int_0^{L/2} \frac{\left(Rx - \frac{qx^2}{2}\right)}{EI}\text{d}x + \int_{L/2}^{L} \frac{\left[Rx - \frac{qL}{2}\left(x - \frac{L}{4}\right) - P\left(x - \frac{L}{2}\right)\right]}{2EI}\text{d}x$$

$$= \frac{(5R - P)L^2}{16EI} - \frac{qL^3}{12EI} = \frac{7qL^3}{192EI} \circlearrowleft$$

Problem 14: The integral frame BCD is of circular cross-section and is loaded at B by three forces in the *x, y, z* directions as shown. Find the net displacement of B.

Further, show that the strain energy contribution due to axial loading may be neglected without much loss of accuracy.

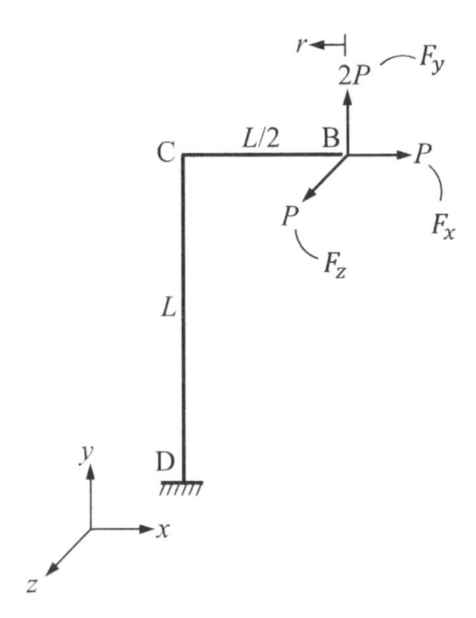

Solution steps:

(1) The three displacement components of the end B are required, and towards that purpose, the applied loads in the three directions are designated as F_x (=P), F_y (=$2P$), and F_z (=P), respectively. (One should note that the net displacement of B does not occur in the same direction as the resultant load.)

(2) The segment BC is subjected to the following loading actions:
- Uniform axial loading of F_x along its length.
- Bending moment due to the combined action of F_y and F_z as given by $\sqrt{\left(F_y^2 + F_z^2\right)}.r$, where r is a local coordinate from B to C as shown.

The corresponding components of the strain energy are

$$U_{\text{bar BC}} = \frac{F_x^2\left(\frac{L}{2}\right)}{2AE}, \quad U_{\text{beam BC}} = \int_0^{\frac{L}{2}} \frac{\left(F_y^2 + F_z^2\right)r^2}{2EI} dr$$

(3) Considering a free body diagram of segment CD as shown, and noting that the bending moment $\frac{F_yL}{2}$ and the torque $\frac{F_zL}{2}$ arise when the applied loads are shifted from B to C, one can identify the following loading actions:
- Uniform axial loading of F_y along the length CD.
- Bending moments of $\left(F_xs - \frac{F_yL}{2}\right)$ in the x–y plane and F_zs in the y–z plane, where s is a local coordinate from C as shown. Thus, the net bending moment distribution is given by $\sqrt{\left(F_xs - \frac{F_yL}{2}\right)^2 + (F_zs)^2}$
- Uniform torque of $\frac{F_zL}{2}$ along the length CD.

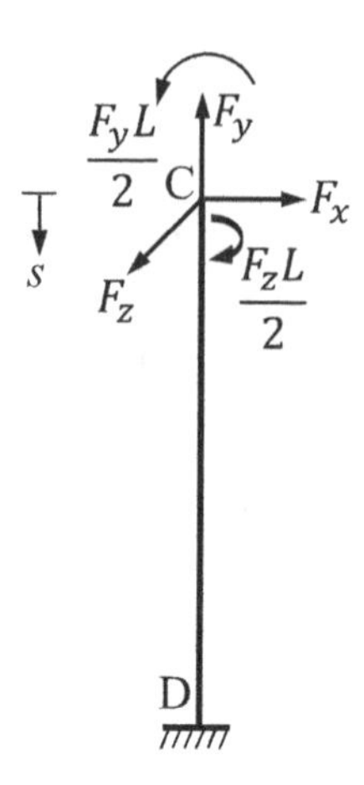

The corresponding strain energy components are

$$U_{\text{bar CD}} = \frac{F_y^2 L}{2AE}, \quad U_{\text{shaft CD}} = \frac{(F_z L/2)^2 L}{2GJ}$$

$$U_{\text{beam CD}} = \int_0^L \frac{\left[(F_x s - F_y L/2)^2 + (F_z s)^2\right]}{2EI} \mathrm{d}s$$

(4) The total strain energy U of the frame BCD is the sum of all the components given above, and its derivatives with respect to F_x, F_y, and F_z yield the three displacement components of B as given below.

$$u_B = \frac{\partial U}{\partial F_x} = \frac{F_x L}{2AE} + \int_0^L \frac{\left(F_x s - \frac{F_y L}{2}\right)s}{EI} \mathrm{d}s$$

$$= \frac{F_x L}{2AE} + \frac{F_x L^3}{3EI} - \frac{F_y L^3}{4EI} = \frac{PL}{2AE} - \frac{PL^3}{6EI}$$

$$v_B = \frac{\partial U}{\partial F_y} = \int_0^{\frac{L}{2}} \left(\frac{F_y r^2}{EI}\right) \mathrm{d}r + \frac{F_y L}{AE} - \int_0^L \frac{\left(F_x s - \frac{F_y L}{2}\right)L}{2EI} \mathrm{d}s$$

$$= \frac{F_y L^3}{24EI} + \frac{F_y L}{AE} - \frac{F_x L^3}{4EI} + \frac{F_y L^3}{4EI} = \frac{2PL}{AE} + \frac{PL^3}{3EI}$$

$$w_B = \frac{\partial U}{\partial F_z} = \int_0^{\frac{L}{2}} \left(\frac{F_z r^2}{EI}\right) \mathrm{d}r + \frac{F_z L^3}{4GJ} + \int_0^L \left(\frac{F_z s^2}{EI}\right) \mathrm{d}s$$

$$= \frac{F_z L^3}{24EI} + \frac{F_z L^3}{4GJ} + \frac{F_z L^3}{3EI} = \frac{3PL^3}{8EI} + \frac{PL^3}{4GJ}$$

The net displacement of B is then obtained as $\sqrt{u_B^2 + v_B^2 + w_B^2}$.

(5) In order to show that the components $U_{\text{bar BC}}$ and $U_{\text{bar CD}}$ may be neglected, consider their contributions to the final displacement components; these are given by $\frac{PL}{2AE}$ to u_B and $\frac{2PL}{AE}$ to v_B. The ratio of either of these to the corresponding remaining term in u_B or v_B turns out to be proportional to I/AL^2 or $\left(\frac{d}{L}\right)^2$, where d is the diameter of the cross-section. Thus, the strain energy contribution due to axial loading is indeed negligible.

(The relative importance of the strain energy contributions due to bending and twisting may be gauged by comparing the two components of w_B; this shows that they are comparable in magnitude and should always be accounted for in problems of combined loading.)

Index

K. Bhaskar and T. K. Varadan, *Strength of Materials*,
https://doi.org/10.1007/978-3-031-06377-0

9783031063763